SpringerBriefs in Applied Sciences and Technology

SpringerBriefs present concise summaries of cutting-edge research and practical applications across a wide spectrum of fields. Featuring compact volumes of 50 to 125 pages, the series covers a range of content from professional to academic.

Typical publications can be:

- A timely report of state-of-the art methods
- An introduction to or a manual for the application of mathematical or computer techniques
- A bridge between new research results, as published in journal articles
- A snapshot of a hot or emerging topic
- An in-depth case study
- A presentation of core concepts that students must understand in order to make independent contributions

SpringerBriefs are characterized by fast, global electronic dissemination, standard publishing contracts, standardized manuscript preparation and formatting guidelines, and expedited production schedules.

On the one hand, **SpringerBriefs in Applied Sciences and Technology** are devoted to the publication of fundamentals and applications within the different classical engineering disciplines as well as in interdisciplinary fields that recently emerged between these areas. On the other hand, as the boundary separating fundamental research and applied technology is more and more dissolving, this series is particularly open to trans-disciplinary topics between fundamental science and engineering.

Indexed by EI-Compendex, SCOPUS and Springerlink.

Manish Kumar Goyal · Shivam Singh

Understanding Atmospheric Rivers Using Machine Learning

 Springer

Manish Kumar Goyal 🆔
Department of Civil Engineering
Indian Institute of Technology Indore
Indore, Madhya Pradesh, India

Shivam Singh 🆔
Department of Civil Engineering
Indian Institute of Technology Indore
Indore, Madhya Pradesh, India

ISSN 2191-530X   ISSN 2191-5318 (electronic)
SpringerBriefs in Applied Sciences and Technology
ISBN 978-3-031-63477-2   ISBN 978-3-031-63478-9 (eBook)
https://doi.org/10.1007/978-3-031-63478-9

This Springer imprint is published by the registered company Springer Nature Switzerland AG
The registered company address is: Gewerbestrasse 11, 6330 Cham, Switzerland

If disposing of this product, please recycle the paper.

Preface

Atmospheric rivers (ARs) are intriguing phenomena that intricately connect climate extremes and the management of water resources. This book delves into the heart of these rivers in the sky, exploring their profound influence on our environment and society. ARs, often termed as the "conveyors of water vapor," play a pivotal role in shaping precipitation patterns, droughts, floods, and water availability across regions. Understanding their characteristics, behaviors, and interactions with large-scale climate oscillations is essential for advancing climate science, water resource management, and disaster risk reduction strategies.

Through a multidisciplinary lens, this book navigates through the complexities of ARs, unraveling their significance in the broader context of climate variability and change. From terrestrial rivers to the far-reaching impacts of climate oscillations, each chapter unfolds a distinct facet of ARs, shedding light on their detection, characterization, impacts, and potential future trajectories. Case studies provide real-world insights into the practical applications of data analytics, machine learning, and innovative technologies in deciphering AR dynamics and enhancing predictive capabilities.

The convergence of science, technology, and innovation in AR research opens new horizons for mitigating risks, improving water management strategies, and fostering resilience in the face of climate challenges. This book is a culmination of collaborative efforts, bringing together experts, researchers, policymakers, and practitioners to delve into the intricate web of atmospheric rivers and their profound implications for our planet's future.

We hope this book serves as a comprehensive guide, igniting curiosity, sparking discussions, and inspiring innovative solutions in the realm of atmospheric rivers and climate extremes.

Indore, India

Manish Kumar Goyal
Shivam Singh

Contents

Chapter 1
Understanding Atmospheric Rivers and Exploring Their Role as Climate Extremes

1.1 Terrestrial Rivers

Rivers symbolize the continuous flow of water through landscapes, serving as vital conduits that shape our world [1]. Carving their way through terrains, they create ecosystems, support biodiversity, and sustain human civilizations. These flowing bodies of water, whether meandering gently or rushing vigorously, play a crucial role in shaping the physical and social landscapes they traverse [15]. Their influence extends beyond geographical boundaries, affecting climates, economies, and cultures along their course.

1.2 Climate Extremes

Climate extremes represent occurrences in weather patterns that significantly depart from typical or average conditions [40, 56]. These deviations manifest in various ways, such as extreme temperatures, abnormal precipitation, prolonged periods of drought or excessive rainfall, intense storms, and heatwaves [6, 10, 13, 32, 44]. They disrupt the usual climatic balance, posing substantial challenges to natural systems, human societies, and built infrastructure. Understanding these events involves delving into their root causes, which can stem from natural variability, human-induced climate change, or a combination of both factors [10, 50].

Beyond observing these anomalies, studying climate extremes entails comprehensive analyses to unravel their impacts and predict their occurrences [18, 52]. Assessing vulnerabilities and risks associated with these extreme weather phenomena is pivotal for devising effective adaptation and mitigation strategies [11, 51]. These strategies aim to minimize the adverse impacts of such events on communities, infrastructure, and the environment [12, 19, 31, 47]. This field of study is vital

M. K. Goyal and S. Singh, *Understanding Atmospheric Rivers Using Machine Learning*, SpringerBriefs in Applied Sciences and Technology, https://doi.org/10.1007/978-3-031-63478-9_1

Table 1.1 Major climate extremes and their mathematical interpretation

S. no	Climate extremes	Interpretation/study approach	Mathematical interpretation
1	Heatwaves	Analyzing prolonged periods of unusually high temperatures	Deviation from the mean temperature over a specific period; for instance, using anomalies or standard deviations from historical temperature data
2	Floods	Studying excessive and rapid inundation of land due to heavy rainfall	Analysis of rainfall intensities, river discharge, and return periods of extreme events through statistical methods like frequency analysis or probability distributions
3	Droughts	Evaluating extended periods of water scarcity or precipitation deficit	Calculating standardized precipitation indices or comparing actual precipitation with long-term averages to determine severity and duration of drought events
4	Storms	Assessing intense weather events characterized by strong winds and heavy rain	Measurement of wind speed, atmospheric pressure, and rainfall intensity using meteorological data or formulas like Beaufort scale for wind speed classification

for enhancing resilience against climatic uncertainties, ensuring preparedness, and fostering sustainable development in the face of an evolving climate.

Analyzing climate extremes involves employing sophisticated techniques, statistical tools, and models to scrutinize historical data and patterns [7, 14, 16, 43, 46]. The primary objective is to comprehend the drivers, frequencies, and impacts of these extreme events. Understanding the mechanisms behind these deviations aids in formulating predictive models and refining our ability to forecast and prepare for extreme weather conditions [4, 16, 43]. By unraveling the complexities of these events, we gain insights crucial for adaptation, resilience building, and disaster management (Table 1.1).

1.3 Atmospheric Rivers: Rivers in the Sky

ARs represent narrow and elongated corridors of moisture-laden air moving within the atmosphere. These concentrated channels of water vapor transport vast amounts of moisture across regions, significantly impacting precipitation patterns (Fig. 1.1). They are integral to the occurrence of intense precipitation events, contributing substantially to the water cycle and precipitation distribution globally.

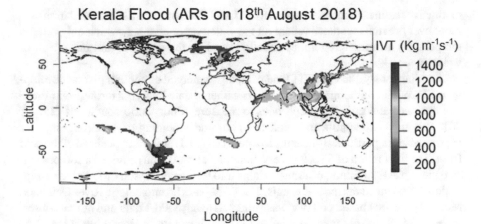

Fig. 1.1 Detected ARs from a global analysis with IVT magnitude on August 18, 2018. The intensified AR over southern India caused an extreme flood in Kerala (India) in August 2018

1.3.1 Historical Background

The understanding of ARs has evolved over time, with significant contributions from researchers in different decades and the integration of various scientific disciplines.

Newell et al. [28] conducted a groundbreaking study on "tropospheric rivers" narrow plumes of strong water vapor transport detected in the atmosphere. Although the term "Atmospheric Rivers" was not coined until later, their work laid the foundation for it. The study by [58] built upon these findings, coining the term "Atmospheric River" and establishing objective techniques for identifying them. They found that ARs were connected to the cold fronts of mid-latitude cyclones and played a major role in global water vapor transport.

By 1998, notable advancements had been achieved in the emerging field of ARs, encompassing the following key developments:

1 Zhu & Newell's 1998 study significantly contributed to the progression of our comprehension of ARs.
2 During the 1997–1998 California Land-Falling Jets (CALJET) field campaign, aircraft were used to provide insightful in situ data for understanding ARs.
3 Availability of an adequate number of polar-orbiting SSM/I satellites facilitated the creation of a nearly synoptic view of integrated water vapor over global oceans.
4 The release of the reanalysis data by NCEP/NCAR for AR characterization.

During the 2000s, a groundbreaking advancement in AR research occurred with the introduction of satellite-based water vapor imagery utilizing the Special Sensor Microwave Imager/Sounder (SSM/I). This technology represented a significant milestone, enabling meteorologists and researchers to analyze Integrated Water Vapor (IWV) in near real time. The use of this imagery provided visualizations of the distinctive features associated with ARs [35, 57]. This marked a substantial improvement over previous methods reliant on the water vapor channel, which only offered temperature information without directly revealing the presence and movement of water vapor.

Simultaneously, the CALJET field campaign emerged as another pivotal development. Through this campaign, researchers gathered data by flying through extratropical cyclones and utilizing dropsondes to measure atmospheric conditions [25, 35, 36]. This approach significantly enhanced our understanding of the strongest water vapor transport, particularly near the low-level jet (LLJ) and along or ahead of the cold front. The integration of SSM/I observations and aircraft data provided a comprehensive view of AR structure, offering insights into the heavy coastal orographic precipitation experienced during AR events. The IWV-based Atmospheric River Detection Method (ARDM) based on IWV was validated using SSM/I data and aircraft observations [34, 57]. This method involved identifying ARs by searching for regions with large IWV meeting specific criteria, such as length, width, and IWV threshold, aligning with in situ observations and reanalysis data. While IWV remained a valuable metric, the focus shifted toward Integrated Water Vapor Transport (IVT), which incorporates wind information. IVT has become the preferred metric due to the accessibility of in situ data observation systems and high-quality atmospheric reanalysis products [8, 21, 22, 27, 47–49].

During the 2000s, the exploration of ARs experienced a surge in research activity. Technological advancements, notably the utilization of the SSM/I, enabled more thorough observations and documentation of ARs. Scientists directed their focus toward understanding the climatology, influence on precipitation events, and hydroclimate impact, particularly in the western United States [3, 41, 42]. It became evident that ARs played a significant role in triggering extreme weather events in specific regions, posing distinctive challenges. In the following decade, the study of ARs expanded its scope, intertwining with diverse disciplines such as hydro-climatology, ecology, and atmospheric aerosols [2, 37, 39, 54]. Collaborative efforts across different fields contributed to a holistic understanding of ARs.

At present, the community engaged in studying ARs encompasses contributors from various disciplines, spanning meteorology hydro-climatology, civil engineering, palaeoclimatology, and polar science [9, 23, 30, 33]. Numerous algorithms have been developed to detect and track AR conditions, identifying ARs at both regional and global scales. Witnessing the impact of ARs and a growing global interest in further exploring these findings, a collaborative initiative, the Atmospheric River Tracking Method Intercomparison Project (ARTMIP), was initiated in 2017 [38, 41, 45]. This project aims to quantify uncertainties arising from different AR detection methods. Studies within ARTMIP have revealed variations in AR counts between methodologies, underscoring the sensitivity of detection methods to geometric criteria.

1.3.2 Mechanism Behind Formation and Intensification of ARs

Research on extreme winter precipitation events in mid-latitude regions over land consistently reveals a strong association with extratropical cyclones [26]. These episodes of heavy precipitation frequently happen when warm, humid air from a cyclone's warm zone comes into contact with elevated terrain, as is the case along the western coasts of the United States [55]. This interaction results in a substantial increase in precipitation depending on the size and shape of AR and the angle of interaction [20]. According to Bader and Roach (1977), these episodes can be described by the interaction of mesoscale and large-scale synoptic processes within a particular orographic precipitation enhancement mechanism called as the "seeder-feeder" mechanism. This feeder cloud typically caps a hill or a small mountain. This phenomenon results in more significant precipitation occurring on the elevated terrain beneath the cap cloud compared to nearby flatlands. Importantly, this mechanism does not significantly alter the overall precipitation associated with the cyclone but serves to redistribute it, concentrating it over regions with elevated terrain [27, 53]. The efficacy of this process is contingent upon the presence of a sufficiently strong, low-level moist airflow to preserve the cloud water content within the orographic feeder cloud and ensure a continuous provision of precipitation from the seeder cloud.

1.3.3 Understanding the Impact of ARs

The study has been carried out to identify the significant impacts of landfalling ARs. The contributions of precipitation driven by ARs can also provide insights in recognizing vulnerable regions for AR associated with floodings. Some case studies with extreme flooding resulted from heavy precipitation have been analyzed in connection with ARs. The case studies were also analyzed to validate the algorithm used for the characterization of ARs. We have found that the heavy precipitation in Mumbai City of India on July 2005, Quebec Province of Canada on May 2017, and Kerala State of India on August 2018 were concurrent with consistent presence of AR with high intensity over the region during that time.

1.4 Case Studies

(a) Kerala Flood (August 2018)

The Kerala floods in August 2018 were a devastating natural disaster that occurred in the southern Kerala, India. The state experienced unusually extreme precipitation during the monsoon season, leading to severe flooding and landslides in many parts of the state (Figs. 1.2 and 1.3). The floods were triggered by the Southwest Monsoon, which arrived in Kerala in early June 2018. However, it significantly escalated in the 2nd week of August, resulting in continuous rainfall and extensive flooding. According to reports from the Central Water Commission of India (CWC) in 2018, the state received 2394.4 mm of rainfall between June 1 and August 19, 2018, exceeding the anticipated 1649.5 mm. The most significant rainfall occurred during the first 3 weeks of August, with the state receiving approximately three times the usual precipitation for that period [5]. This intense precipitation led to the overflowing of numerous rivers and tributaries, causing flooding in low-lying areas and displacing thousands of residents from their homes.

The torrential rains and flooding had a catastrophic impact on Kerala's infrastructure, agriculture, and economy. The state witnessed massive destruction of homes, buildings, roads, bridges, and other critical infrastructure. Numerous villages and towns were entirely submerged, and both communication and transportation systems faced severe disruptions. Thirteen out of the fourteen districts in Kerala experienced extensive flooding due to heavy rainfall from August 13 to 17, 2018 [24]. The 2018 Kerala flood resulted in over 500 fatalities and the inundation of 775 townships, as reported by Revenue and Disaster Management in 2018. Since ARs can carry huge amounts of water and lead to extreme precipitation in a shorter span. To understand the reason behind the extreme rainfall on 13–17 August, we explored the IVT characteristics and possibility of conditions for an intense AR (Fig. 1.2).

These intense IVT characteristics were analyzed with defined thresholds for identifying ARs during this period of August 12–20, 2018, and at multiple steps, ARs were identified with intense IVT characteristics (Fig. 1.3). Two such timesteps on August 16, 2018 at 00:00 and 06:00 were presented in Fig. 1.4.

The analysis shows that the extreme precipitation experienced by the Indian state of Kerala on August 13–17, 2018 was driven by AR.

(b) Quebec, Canada Flood (May 2017)

According to reports, a total of 5,371 residences were flooded, approximately 4,066 people were forced to evacuate their homes, and around 261 municipalities in the region were got affected (Lin et al. 2019) (Fig. 1.5). The floods were ranked as the third major weather event of 2017 by the Canadian Meteorological and Oceanographic Society [29]. In response to the severity of the situation, a state of emergency was declared in the cities of Montreal and Laval.

The ARs with intense characteristics were available consistently over the Quebec province during the first week of May 2017 and were potentially responsible for intense precipitation and snowfall during that period (Fig. 1.6).

(c) Mumbai Flood (July 2005)

On July 26, 2005, the city experienced exceptionally heavy rainfall during the monsoon season, leading to severe flooding and widespread chaos [17]. The torrential rain began in the early morning of 26 July and continued for several hours, causing several rivers and drainage systems to overflow. The flooding caused significant damage to buildings, houses, and public infrastructure (Fig. 1.7).

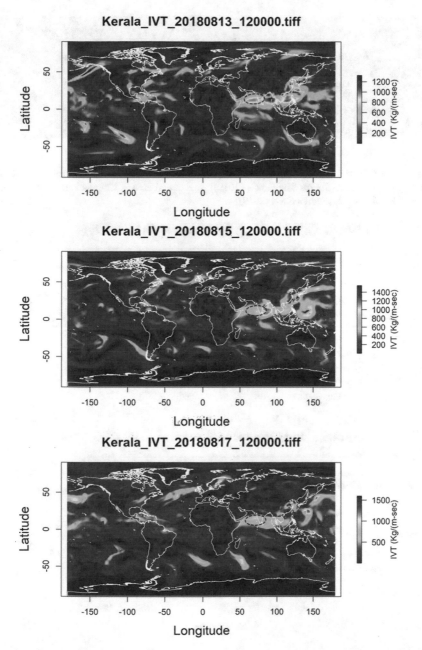

Fig. 1.2 Maps exhibiting the intense IVT characteristics existing in the southern region of India from August 13, 15, and 17 2018 (one map chosen from 6 hourly maps for each day)

        (a)   (b)

Fig. 1.3. **a** An aerial view of a bridge that is the only object above the flood water in Central Kerala on August 19, 2018. [*Source* The Hindu (2019)]; **b** Flooded area in Chengannur in the southern state of Kerala, India, [*Source* CNN (2019)]

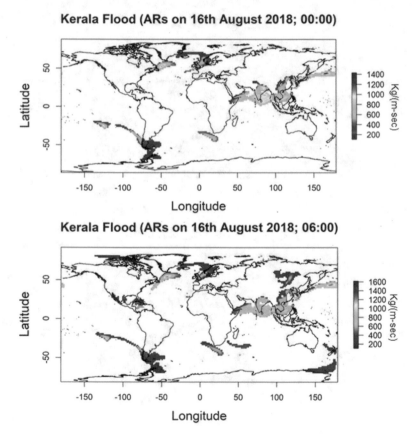

Fig. 1.4 Presence of intense AR over southern regions of India on August 16, 2018 at 00:00 and 06:00

(a)   (b)

Fig. 1.5. a On May 4, 2017, flooded residential street in Gatineau, Quebec, Canada [*source* 61];
b a resident removes belongings from his home in a flooded residential neighborhood in Rigaud,
Quebec, Canada May 7, 2017 [*source* 62]

Mumbai City experienced a standstill due to record-breaking rainfall of 994 mm in
24 h starting from 08:30 on July 26, 2005. Following the events, landslides and flash
floods in the Mumbai municipal area resulted in the loss of at least 419 lives, including
16,000 animals and an extra 216 deaths were attributed to flood-related illnesses
[59]. The consequences also encompassed damage to over 100,000 residential and
commercial structures, along with 30,000 vehicles [60].

The flooding also led to massive economic losses, as businesses suffered from
disruptions and damage to property and infrastructure. From July 23 onward, there
was a noticeable amount of moisture shift on the western coast of India heading
toward Maharashtra, India. These intense IVT characteristics representing potential
ARs lasted for several days in the last week of July and the first week of August
(Fig. 1.8). Due to these intense IVT characteristics concentrated over peninsular
India, an AR with high intensity was consistently identified over the region for a
couple of days, as shown in Fig. 1.9.

The heavy moisture was concentrated over the region in a couple of days starting
from 23 July, and instability caused by that intense moisture representing as an
AR over the coastal city of Maharashtra, Mumbai, would have resulted in extreme
precipitation. Even after this spell of heavy rainfall, it remained consistent with
intense moisture for some more days over the same region with rapidly intensifying
the moisture over a location ahead in the western Pacific Ocean (Fig. 1.8).

These ARs could bring primarily beneficial to primarily hazardous impacts over
a region depending upon the intensity and orientation of AR and the portion of AR
making landfall. The continents of mid-latitudes have been noticed to experience
significant impacts of ARs annually. These ARs also contribute to snowfall in polar
regions. Observing the global impact of ARs, it becomes important to study the
detailed impacts, and drivers of ARs.

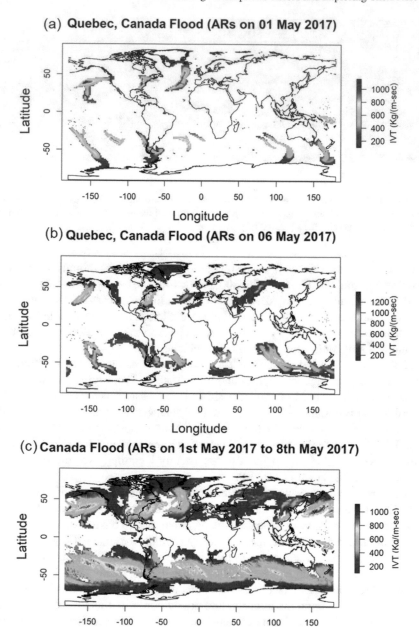

Fig. 1.6 Maps showing the presence of an intense AR (average of all four steps of AR detection for a day from 6 hourly datasets) over the Canadian province of Quebec in 1st week of May 2017, specifically on **a** 1st May, **b** 7th of May, and **c** the average IVT characteristics during the presence of AR over the region from a week from May 1–8, 2017

(a)   (b)

Fig. 1.7 a Commuters wade through floodwaters following intense torrential rains that brought the city of Mumbai to a standstill on July 27, 2005 (AFP photo) [*source* NEWS18 (2005a)]; **b** commuters walk through floodwaters past stranded motor vehicles after heavy torrential rains paralyzed the city of Mumbai, July 27, 2005 [(*source* NEWS18 (2005b)]

1.5 Conclusions

The discussion on climate extremes emphasizes the disruptive nature of deviations from typical weather patterns, underscoring the importance of understanding their root causes. This understanding is crucial for developing effective strategies to cope with and adapt to the challenges posed by climate extremes. Going beyond simple observation, there is a call for comprehensive analyses to uncover the impacts and predict the occurrences of these extremes. This foundational knowledge is essential for evaluating vulnerabilities, managing risks, and fostering resilience in the face of a changing climate.

ARs are one such climate extreme, characterized as narrow passages of moisture-laden air with profound implications for global precipitation patterns. Exploring the historical background of AR research offers a temporal perspective on its evolution, highlighting pivotal moments in the late twentieth century and substantial technological advancements during the 2000s. The "seeder-feeder" mechanism emerges as a critical factor in understanding episodes of heavy precipitation associated with ARs. To understand the broader impacts of landfalling ARs, a meticulous examination of detailed case studies from regions such as Kerala in August 2018, Quebec in May 2017, and Mumbai in July 2005 is undertaken. These real-world examples serve as poignant reminders of the severe consequences of AR-associated floods, emphasizing the vital role of robust algorithms and early warning systems in mitigating such events. The geographical diversity portrayed by these case studies underscores the global significance of ARs and their diverse impacts across different regions.

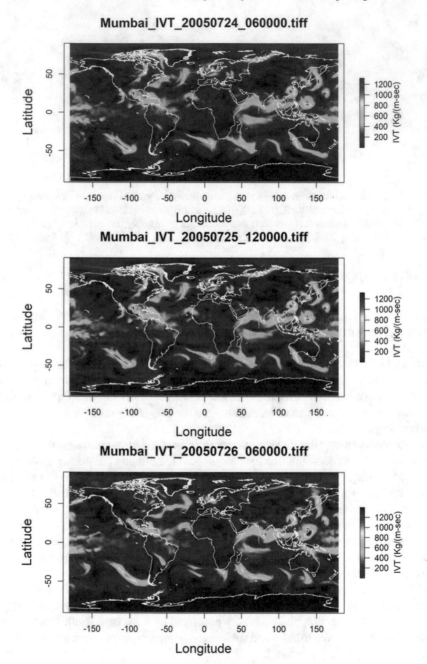

Fig. 1.8 Maps showing the intense IVT characteristics existing heading toward India from July 24, 25, 26, 2005 (one map chosen from 6 hourly maps for each day)

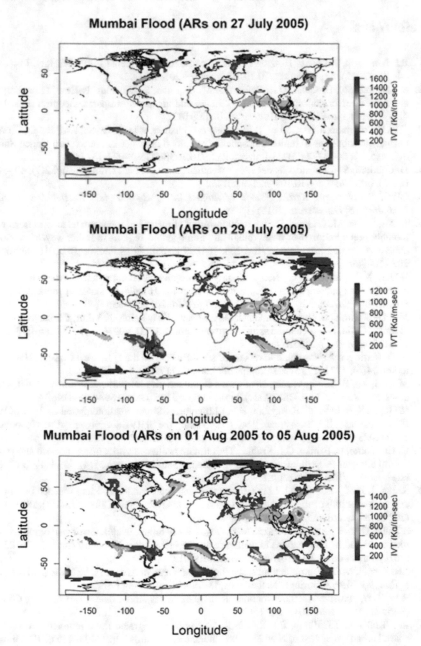

Fig. 1.9 Presence of intense AR over southern regions of India from July 27 to 29, 2005, and average AR characteristics on August 1–5, 2005

References

1. P.J. Ashworth, J. Lewin, How do big rivers come to be different? Earth Sci. Rev. **114**(1–2), 84–107 (2012). https://doi.org/10.1016/j.earscirev.2012.05.003
2. S. Chakraborty, B. Guan, D.E. Waliser, A.M. da Silva, S. Uluatam, P. Hess, Extending the atmospheric river concept to aerosols: climate and air quality impacts. Geophys. Res. Lett. **48**(9) (2021). https://doi.org/10.1029/2020GL091827
3. T.W. Corringham, J. McCarthy, T. Shulgina, A. Gershunov, D.R. Cayan, F.M. Ralph, Climate change contributions to future atmospheric river flood damages in the western United States. Sci. Rep. **12**(1), 13747 (2022). https://doi.org/10.1038/s41598-022-15474-2
4. D. Coumou, S. Rahmstorf, A decade of weather extremes. Nat. Clim. Chang. **2**(7), 491–496 (2012). https://doi.org/10.1038/nclimate1452
5. CWC, *Study report Kerala Floods of August 2018, Hydrological Studies Organisation Hydrology (S) Directorate* (2018)
6. J. Das, S. Jha, M.K. Goyal, On the relationship of climatic and monsoon teleconnections with monthly precipitation over meteorologically homogenous regions in India: wavelet & global coherence approaches. Atmos. Res. **238**, 104889 (2020). https://doi.org/10.1016/j.atmosres.2020.104889
7. J. Das, S. Jha, M.K. Goyal, Non-stationary and copula-based approach to assess the drought characteristics encompassing climate indices over the Himalayan states in India. J. Hydrol. **580**, 124356 (2020). https://doi.org/10.1016/j.jhydrol.2019.124356
8. M.J. DeFlorio, D.E. Waliser, B. Guan, D.A. Lavers, F.M. Ralph, F. Vitart, Global assessment of atmospheric river prediction skill. J. Hydrometeorol. **19**(2), 409–426 (2018). https://doi.org/10.1175/JHM-D-17-0135.1
9. M.D. Dettinger, Atmospheric Rivers as Drought Busters on the U.S. West Coast. J. Hydrometeorol. **14**(6), 1721–1732 (2013). https://doi.org/10.1175/JHM-D-13-02.1
10. M.K. Goyal, S. Singh, V. Jain, Heat waves characteristics intensification across Indian smart cities. Sci. Rep. **13**(1), 14786 (2023). https://doi.org/10.1038/s41598-023-41968-8
11. V. Gupta, S. Rakkasagi, S. Rajpoot, H.S. El Imanni, S. Singh, Spatiotemporal analysis of Imja Lake to estimate the downstream flood hazard using the SHIVEK approach. Acta Geophysica 1–12 (2023)
12. U. Huiskamp, B. Brinke, G.J. Kramer, The climate resilience cycle: using scenario analysis to inform climate-resilient business strategies. Bus. Strateg. Environ. **31**(4), 1763–1775 (2022). https://doi.org/10.1002/bse.2982
13. S. Jha, J. Das, M.K. Goyal, Low frequency global-scale modes and its influence on rainfall extremes over India: nonstationary and uncertainty analysis. Int. J. Climatol. **41**(3), 1873–1888 (2021). https://doi.org/10.1002/joc.6935
14. S. Jha, J. Das, A. Sharma, B. Hazra, M.K. Goyal, Probabilistic evaluation of vegetation drought likelihood and its implications to resilience across India. Glob. Planetary Change **176**, 23–35 (2019). https://doi.org/10.1016/j.gloplacha.2019.01.014
15. J.R. Karr, E.W. Chu, Introduction: sustaining living rivers, in *Assessing the Ecological Integrity of Running Waters* (Springer Netherlands, Dordrecht), pp. 1–14 (2000)
16. R.W. Katz, Statistical methods for nonstationary extremes, in *Extremes in a Changing Climate* (Springer, Dordrecht, 2013), pp 15–37
17. A.T. Kulkarni, T.I. Eldho, E.P. Rao, B.K. Mohan, An integrated flood inundation model for coastal urban watershed of Navi Mumbai, India. Nat. Hazards **73**(2), 403–425 (2014). https://doi.org/10.1007/s11069-014-1079-6
18. N. Kumar, P. Patel, S. Singh, M.K. Goyal, Understanding non-stationarity of hydroclimatic extremes and resilience in Peninsular catchments, India. Sci. Rep. **13**(1), 12524 (2023). https://doi.org/10.1038/s41598-023-38771-w
19. N. Kumar, V. Poonia, B.B. Gupta, M.K. Goyal, A novel framework for risk assessment and resilience of critical infrastructure towards climate change. Technol. Forecast. Soc. Chang. **165**, 120532 (2021). https://doi.org/10.1016/j.techfore.2020.120532

20. J. Liang, Y. Yong, M.K. Hawcroft, Long-term trends in atmospheric rivers over East Asia. Clim. Dyn. **60**(3–4), 643–666 (2023). https://doi.org/10.1007/s00382-022-06339-5

21. R.V. Lyngwa, M.A. Nayak, Atmospheric river linked to extreme rainfall events over Kerala in August 2018. Atmos. Res. **253**, 105488 (2021). https://doi.org/10.1016/j.atmosres.2021.105488

22. S. Meghani, S. Singh, N. Kumar, M.K. Goyal, Predicting the spatiotemporal characteristics of atmospheric rivers: a novel data-driven approach. Glob. Planet. Change **231**, 104295 (2023). https://doi.org/10.1016/j.gloplacha.2023.104295

23. A.C. Michaelis, A. Gershunov, A. Weyant, M.A. Fish, T. Shulgina, F.M. Ralph, Atmospheric river precipitation enhanced by climate change: a case study of the storm that contributed to California's Oroville dam crisis. Earth's Future **10**(3) (2022). https://doi.org/10.1029/2021EF002537

24. V. Mishra, H.L. Shah, Hydroclimatological perspective of the Kerala flood of 2018. J. Geol. Soc. India **92**(5), 645–650 (2018). https://doi.org/10.1007/s12594-018-1079-3

25. P.J. Neiman, F.M. Ralph, A.B. White, D.E. Kingsmill, P.O.G. Persson, The statistical relationship between upslope flow and rainfall in California's coastal mountains: observations during CALJET. Monthly weather review. Am. Meteorol. Soc. **130**(6), 1468–1492 (2002). https://doi.org/10.1175/1520-0493(2002)130%3c1468:TSRBUF%3e2.0.CO;2

26. P.J. Neiman, F.M. Ralph, G.A. Wick, J.D. Lundquist, M.D. Dettinger, Meteorological characteristics and overland precipitation impacts of atmospheric rivers affecting the West Coast of North America based on eight years of SSM/I satellite observations. J. Hydrometeorol. Am. Meteorol. Soc. **9**(1), 22–47 (2008). https://doi.org/10.1175/2007JHM855.1

27. P.J. Neiman, L.J. Schick, F.M. Ralph, M. Hughes, G.A. Wick, Flooding in Western Washington: the connection to atmospheric rivers. J. Hydrometeorol. **12**(6), 1337–1358 (2011). https://doi.org/10.1175/2011JHM1358.1

28. R.E. Newell, N.E. Newell, Y. Zhu, C. Scott, Tropospheric rivers?—a pilot study. Geophys. Res. Lett. **19**(24), 2401–2404 (1992). https://doi.org/10.1029/92GL02916

29. I. Olthof, N. Svacina, Testing urban flood mapping approaches from satellite and in-situ data collected during 2017 and 2019 events in eastern Canada. Remote Sens. **12**(19), 3141 (2020). https://doi.org/10.3390/rs12193141

30. A.E. Payne, M. E. Demory, L.R. Leung, A.M. Ramos, C.A. Shields, J.J. Rutz, N. Siler, G. Villarini, A. Hall, F.M. Ralph, Responses and impacts of atmospheric rivers to climate change. Nat. Rev. Earth Environ. **1**(3), 143–157 (2020). https://doi.org/10.1038/s43017-020-0030-5

31. V. Poonia, M.K. Goyal, B.B. Gupta, A.K. Gupta, S. Jha, J. Das, Drought occurrence in Different River Basins of India and blockchain technology based framework for disaster management. J. Clean. Prod. **312**, 127737 (2021). https://doi.org/10.1016/j.jclepro.2021.127737

32. S. Rakkasagi, M.K. Goyal, S. Jha, Evaluating the future risk of coastal Ramsar wetlands in India to extreme rainfalls using fuzzy logic. J. Hydrol. **632**, 130869 (2024). https://doi.org/10.1016/j.jhydrol.2024.130869

33. F.M. Ralph, M. Dettinger, D. Lavers, I.V. Gorodetskaya, A. Martin, M. Viale, A.B. White, N. Oakley, J. Rutz, J.R. Spackman, H. Wernli, J. Cordeira, Atmospheric rivers emerge as a global science and applications focus. Bull. Am. Meteor. Soc. **98**(9), 1969–1973 (2017). https://doi.org/10.1175/BAMS-D-16-0262.1

34. F.M. Ralph, M.D. Dettinger, Storms, floods, and the science of atmospheric rivers. Eos Trans. Am. Geophys. Union **92**(32), 265–266 (2011). https://doi.org/10.1029/2011EO320001

35. F.M. Ralph, P.J. Neiman, R. Rotunno, Dropsonde observations in low-level jets over the northeastern pacific ocean from CALJET-1998 and PACJET-2001: mean vertical-profile and atmospheric-river characteristics. Monthly Weather Rev. Am. Meteorol. Soc. Boston MA, USA, **133**(4), 889–910 (2005). https://doi.org/10.1175/MWR2896.1

36. F.M. Ralph, P.J. Neiman, G.A. Wick, Satellite and CALJET aircraft observations of atmospheric rivers over the eastern north pacific ocean during the winter of 1997/98. Monthly Weather Rev. Am. Meteorol. Soc. **132**(7), 1721–1745 (2004). https://doi.org/10.1175/1520-0493(2004)132%3c1721:SACAOO%3e2.0.CO;2

37. F.M. Ralph, K.A. Prather, D. Cayan, J.R. Spackman, P. DeMott, M. Dettinger, C. Fairall, R. Leung, D. Rosenfeld, S. Rutledge, D. Waliser, A.B. White, J. Cordeira, A. Martin, J. Helly, J. Intrieri, CalWater field studies designed to quantify the roles of atmospheric rivers and aerosols in modulating U.S. West. Bull. Am. Meteorol. Soc. **97**(7), 1209–1228 (2016). https://doi.org/10.1175/BAMS-D-14-00043.1

38. F.M. Ralph, A.M. Wilson, T. Shulgina, B. Kawzenuk, S. Sellars, J.J. Rutz, M.A. Lamjiri, E.A. Barnes, A. Gershunov, B. Guan, K.M. Nardi, T. Osborne, G.A. Wick, ARTMIP-early start comparison of atmospheric river detection tools: how many atmospheric rivers hit northern California's Russian River watershed? Clim. Dyn. **52**(7–8), 4973–4994 (2019). https://doi.org/10.1007/s00382-018-4427-5

39. K.S. Rautela, S. Singh, M.K. Goyal, Characterizing the spatio-temporal distribution, detection, and prediction of aerosol atmospheric rivers on a global scale. J. Environ. Manage. **351**, 119675 (2024). https://doi.org/10.1016/j.jenvman.2023.119675

40. M. Rummukainen, Changes in climate and weather extremes in the 21st century. Wiley Interdiscip. Rev. Clim. Change. **3**(2), 115–129 (2012)

41. J.J. Rutz, C.A. Shields, J.M. Lora, A.E. Payne, B. Guan, P. Ullrich, T. O'Brien, L.R. Leung, F.M. Ralph, M. Wehner, S. Brands, A. Collow, N. Goldenson, I. Gorodetskaya, H. Griffith, K. Kashinath, B. Kawzenuk, H. Krishnan, V. Kurlin, D. Lavers, G. Magnusdottir, K. Mahoney, E. McClenny, G. Muszynski, P.D. Nguyen, M. Prabhat, Y. Qian, A.M. Ramos, C. Sarangi, S. Sellars, T. Shulgina, R. Tome, D. Waliser, D. Walton, G. Wick, A.M. Wilson, M. Viale, The atmospheric river tracking method intercomparison project (ARTMIP): quantifying uncertainties in atmospheric river climatology. J. Geophys. Res. Atmos. **124**(24), 13777–13802 (2019). https://doi.org/10.1029/2019JD030936

42. J.J. Rutz, W.J. Steenburgh, F.M. Ralph, The inland penetration of atmospheric rivers over western North America: a Lagrangian analysis. Monthly weather review. Am. Meteorol. Soc. **143**(5), 1924–1944 (2015). https://doi.org/10.1175/MWR-D-14-00288.1

43. S.B. Seo, Y.-O. Kim, Y. Kim,H.-I. Eum, Selecting climate change scenarios for regional hydrologic impact studies based on climate extremes indices. Clim. Dyn. **52**(3–4), 1595–1611 (2019). https://doi.org/10.1007/s00382-018-4210-7

44. A. Sharma, M.K. Goyal, Assessment of drought trend and variability in India using wavelet transform. Hydrol. Sci. J. **65**(9), 1539–1554 (2020). https://doi.org/10.1080/02626667.2020.1754422

45. C.A. Shields, Atmospheric river tracking method intercomparison project (ARTMIP): project goals and experimental design. Geosci. Model Dev. **11**, 2455–2474 (2018). https://doi.org/10.5194/gmd-11-2455-2018

46. G.M.K. Shivam, A.K. Sarma, Analysis of the change in temperature trends in Subansiri River basin for RCP scenarios using CMIP5 datasets. Theoret. Appl. Climatol. **129**(3–4), 1175–1187 (2017). https://doi.org/10.1007/s00704-016-1842-6

47. S. Singh, M.K. Goyal, Enhancing climate resilience in businesses: The role of artificial intelligence. J. Clean. Prod. 138228 (2023)

48. S. Singh, M.K. Goyal, An innovative approach to predict atmospheric rivers: exploring convolutional autoencoder. Atmos. Res. **289**, 106754 (2023)

49. S. Singh, M.K. Goyal, S. Jha, Role of large-scale climate oscillations in precipitation extremes associated with atmospheric rivers: nonstationary framework. Hydrol. Sci. J. **68**(3), 395–411 (2023)

50. S. Singh, N. Kumar, M.K. Goyal, S. Jha, Relative influence of ENSO, IOD, and AMO over spatiotemporal variability of hydroclimatic extremes in Narmada basin, India. AQUA Water Infrastruct. Ecosyst. Soc. **72**(4), 520–539 (2023). https://doi.org/10.2166/aqua.2023.219

51. S. Singh, B. Prasad, H.L. Tiwari, Sedimentation analysis for a reservoir using remote sensing and GIS techniques. ISH J. Hydraulic Eng. **29**(1), 71–79 (2023)

52. S. Singh, A. Yadav, G.M. Kumar, Univariate and bivariate spatiotemporal characteristics of heat waves and relative influence of large-scale climate oscillations over India. J. Hydrol. **628**, 130596 (2024). https://doi.org/10.1016/j.jhydrol.2023.130596

53. B.L. Smith, S.E. Yuter, P.J. Neiman, D.E. Kingsmill, Water vapor fluxes and orographic precipitation over Northern California associated with a landfalling atmospheric river. Mon. Weather Rev. **138**(1), 74–100 (2010). https://doi.org/10.1175/2009MWR2939.1

54. S. Verma, M.V. Ramana, R. Kumar, Atmospheric rivers fueling the intensification of fog and haze over Indo-Gangetic Plains. Sci. Rep. **12**(1), 5139 (2022). https://doi.org/10.1038/s41598-022-09206-9

55. M. Viale, R. Valenzuela, R.D. Garreaud, F.M. Ralph, Impacts of atmospheric rivers on precipitation in Southern South America. J. Hydrometeorol. Am. Meteorol. Soc. **19**(10), 1671–1687 (2018). https://doi.org/10.1175/JHM-D-18-0006.1

56. N. Vivekanandan, S. Singh, M.K. Goyal, Comparison of probability distributions for extreme value analysis and predicting monthly rainfall pattern using Bayesian regularized ANN 271–294 (2023)

57. G.A. Wick, Y.-H. Kuo, F.M. Ralph, T.-K. Wee, P.J. Neiman, Intercomparison of integrated water vapor retrievals from SSM/I and COSMIC. Geophys. Res. Lett. **35**(21), L21805 (2008). https://doi.org/10.1029/2008GL035126

58. Y. Zhu, R.E. Newell, Atmospheric rivers and bombs. Geophys. Res. Lett. **21**(18), 1999–2002 (1994). https://doi.org/10.1029/94GL01710

59. R.B. Bhagat, M. Guha, A. Chattopadhyay, Mumbai after 26/7 Deluge: Issues and Concerns in Urban Planning. Population and Environment, **27**(4), 337–349 (2006). https://doi.org/10.1007/s11111-006-0028-z

60. K. Gupta, Urban flood resilience planning and management and lessons for the future: a case study of Mumbai, India. Urban Water Journal, **4**(3), 183–194 (2007). https://doi.org/10.1080/15730620701464141

61. MACLEAN'S. A woman paddles a kayak past an abandoned car on a flooded residential street in Gatineau, Quebec, Canada. (2017a, May 4).

62. MACLEAN'S. A resident removes belongings from his home in a flooded residential neighbourhood in Rigaud, Quebec, Canada. (2017b, May 7)

Chapter 2
Characterization and Impacts of Atmospheric Rivers

2.1 Introduction

Approximately 90% of atmospheric moisture in mid-latitudes is conveyed by these ARs on average [89]. Scientifically recognized as atmospheric moisture transports or moisture conveyor belts [21], these narrow corridors with high moisture content dynamically alter their courses, obtaining moisture through local convergence, evaporation, and, at times, from distant source regions in the tropics and subtropics [19, 84]. Laden with substantial vapor, these moisture-laden ARs create conditions conducive to induced precipitation through interactions with topography [67]. As a result, landfalling ARs can bring about significant hydrological impacts, including the supply of water, precipitation extremes, and related hazards. Floods are among the most disastrous extreme weather events and have caused devastating damage to infrastructure, the natural environment and human life around the world [13, 14, 23, 35, 59, 72]. Absence of ARs over a particular region in significant numbers in a year could invite drought situations. Droughts also cause several adverse impacts on the health of human beings, animals, vegetations, etc. [24, 34, 39, 57, 77, 81].

Most AR-associated precipitation and flooding events are observed in coastal regions of mid-latitudes. ARs contribute to around 20–30% of annual precipitation on the west coast of the U.S. [40], and approximately 14–44% of total precipitation in warm season in East Asia [36]. Globally, four to five ARs, on average, exist at any given time in each hemisphere, with a higher frequency over the west coasts of South and North America, South Africa, Europe, New Zealand, Australia, and Japan. The polar regions are also affected, with ARs contributing to snowfall [2, 86]. The socio-economic impacts of ARs extend to various western coastal regions worldwide, being responsible for numerous extreme precipitation events.

ARs are renowned for generating heavy precipitation, often resulting in extreme rainfall events and subsequent flooding [15, 47, 53]. The concentrated moisture transport within ARs can lead to intense rainfall rates over short durations, overwhelming drainage systems and causing flash floods, riverine flooding, and landslides [60, 82,

83, 89]. These impacts encompass property damage, infrastructure disruption, and threats to human safety. Of the ten deadliest floods in India from 1985 to 2020, seven were linked to ARs [45]. In 2017, an AR hitting California caused an economic loss of over $1 billion, while in 2019, another AR led to severe flooding in northern California, resulting in damages exceeding $100 million [31]. Identifying regions vulnerable to AR-associated flooding is crucial for effective mitigation and preparedness efforts. Observing their impacts at regional and global scales several approaches have been introduced to characterize these ARs in the troposphere.

Between 2005 and 2017, several field campaigns were conducted to directly observe vapor movement through ARs over the northeast Pacific [54]. These campaigns, such as "Ghost Nets" in 2005, "WISPAR" in 2011, "CalWater" in 2014 and 2015, and "AR Recon" in 2016, involved releasing dropsondes from weather reconnaissance aircraft to record various meteorological conditions at multiple pressure levels [65]. Using these measurements, IWV and IVT were computed, offering insights into AR width, strength, and total IVT. The campaigns disclosed that the average AR is approximately 1000 km wide and transports around 5×10^8 kg s^{-1} of water vapor, comparable to substantial river discharges. Observations and composite analysis of dropsonde transects and reanalysis data provided an understanding into the profile of ARs in lower troposphere. ARs intersect the warm front and undergo depletion through precipitation from ascent in the WCB [61]. The majority of IVT is found below 3 km mean sea level (MSL), with a smaller portion occurring in the upper jet.

Studies incorporating reanalysis products alongside aircraft observations expanded the dataset and further examined AR characteristics. By applying the ARDM with a spatially variable threshold based on IVT, nearly 6000 ARs were identified in the northeast Pacific from 1979 to 2016 [49]. The analysis revealed mean AR widths and total IVTs consistent with aircraft observations, strengthening the robustness of the findings. Moisture transport within ARs involves large-scale horizontal moisture convergence, with the majority originating from subtropical and mid-latitude regions [26, 25]. Moisture replenishment occurs along the AR rather than its entire length, influenced by the size and evolution of associated cyclones.

2.2   Observation and Detection

2.2.1   Satellite-Based Observations

Satellite observations were the initial thrust in studying the science of ARs [27, 52, 66, 67, 62, 88]. Satellites offer a global perspective, providing unique insights into various features (availability, extent, and intensity) of AR events. These observations are particularly valuable as they furnish information not attainable from other existing observing systems or numerical models.

Two types of satellite measurements have significantly contributed to advancing our understanding of ARs:

- **Passive Microwave Radiometric Imagery**: The integration of research aircraft data with this technology enabled the inaugural offshore measurements of ARs. Utilizing passive microwave radiometers, retrievals of IWV provide clear and detailed depictions of ARs. While IWV only may not provide all the characteristics of ARs, it becomes a potent proxy when complemented with other information, such as wind data [28, 83].
- **Radio Occultation Measurements**: These measurements have provided useful water vapor profile evidence over oceans, offering unique insights into the vertical features of ARs [87].

Microwave radiometry, specifically the Special Sensor Microwave/Imager (SSM/I), has been instrumental in revitalizing AR research [9, 63, 86]. In contrast to infrared-based measurements, microwave observations can penetrate the clouds often associated with ARs, allowing accurate measurements of IWV. SSM/I imagery has offered a plan-view context, facilitating the detection and study of the elongated nature of ARs. This led to the development of ARDM, utilizing IWV thresholds above 20 mm for length and width requirements to effectively identify ARs [60, 88]. The availability of daily SSM/I measurements has significantly increased the number of ARs that can be studied compared to research aircraft measurements. IWV imageries derived from satellites have been widely employed in AR domain to identify and characterize ARs in various regions, develop climatology, understand hydrologic impacts, validate forecast models and reanalyses, and support public outreach activities [27, 52].

2.2.2 Reanalysis Data

These types of data are widely used in the study of hydrometeorology owing to their extensive spatiotemporal coverage and detailed representation of the climate system [7]. Reanalysis involves the creation of the present state of the Earth by merging forecast and observational data, serving as a valuable tool for investigating large-scale weather phenomena like ARs [3, 13, 12, 54]. The introduction of NCEP–NCAR reanalysis datasets has significantly enhanced the exploration of the impact and climatology of ARs globally. These datasets, spanning from 1948 onward and relying on radiosonde data, have been widely employed for AR detection [3]. Modern reanalysis products utilizing several satellite observations provide hydro-meteorological observations from 1979 onward, enhancing understanding of the global water cycle [41]. Comparisons between reanalysis products and observations, such as aircraft measurements and satellite records, have indicated that AR features are generally well represented in reanalyses, although some errors persist, varying in magnitude among specific reanalysis products. Reanalysis products have proven particularly valuable in AR studies due to their capacity to provide comprehensive information on vector winds, vertical profiles, and the state of the climate system. Reanalysis products have been applied in various AR research applications, including the examination of spatial patterns and vertical structures of AR landfalls, assessment of historical AR activity, development of ARDMs, and evaluation of global weather and climate models [10, 82, 73, 80]. The assimilation of satellite and other observational data in reanalysis allows for a collective contribution of these observations to AR studies.

2.3 AR Identification

Numerous AR identification methods have emerged, employing either IWV or IVT as thresholding criteria [41, 47]. These threshold values may remain constant or vary based on location, season, or climatology. AR identification considers geometric criteria like length, width, aspect ratio, and curvature. Some methods involve tracking ARs either spatially (Eulerian) or by following them as objects (Lagrangian). [16, 71, 90]. Temporal criteria, machine learning techniques, filtering out non-AR phenomena, and choosing between subjective analysis or objective algorithms are additional considerations. This array of choices has led to diverse permutations and a rich literature on AR identification methods.

Initially, ARs were identified based on IWV thresholds utilizing satellite and aircraft observations, laying the groundwork for assessing AR contributions to precipitation (Fig. 2.1a). Subsequently, with the aid of reanalysis data, IVT-based characterization of ARs gained preference over IWV. Objective methods utilizing IVT thresholds were developed, enabling consistent examination of AR characteristics globally (Figs. 2.1 and 2.2).

Recent endeavors have sought to establish a consensus on AR definition and identification methods. The ARTMIP quantifies uncertainties arising from different AR detection methods [68, 72, 78]. Studies within ARTMIP have revealed differences in the total numbers of AR detected between different algorithms, emphasizing the sensitivity of detection methods to geometric criteria. Nevertheless, there is relatively little difference in pertinent AR key attributes, such as frequency, intensity, and duration.

2.4 Global and Local Insights

Initially, the impacts of ARs were predominantly studied along the west coast of North America and Europe [11, 18, 20, 32, 40, 56, 76]. Recent research on ARs has expanded beyond these regions to encompass polar areas, eastern continental and coastal regions, and other areas historically receiving less scientific attention [36, 37, 42, 43, 45, 74, 79] (Fig. 2.3). These advancements, coupled with the necessity

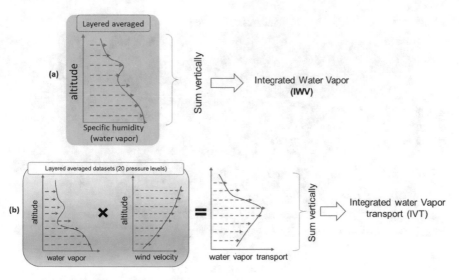

Fig. 2.1 Physical representation of **a** integrated water vapor; **b** integrated water vapor transport for characterizing ARs in the troposphere

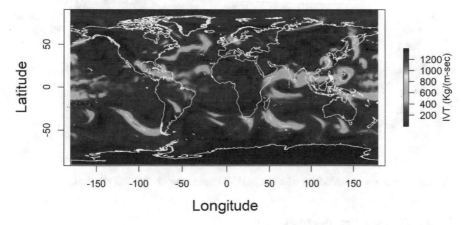

Fig. 2.2 The spatial distribution of IVT (kg m^{-1} s^{-1}) on July 26, 2005, at 06:00 derived from ERA5 (ECMWF) reanalysis data

of benchmarking weather and climate models, underscore the significance of characterizing ARs on a global scale. To meet this objective, various ARDMs suitable for different regions are in development. The ARTMIP outlines the details of these algorithms and provides an opportunity to compare the outcomes derived from these diverse approaches [50, 58, 78].

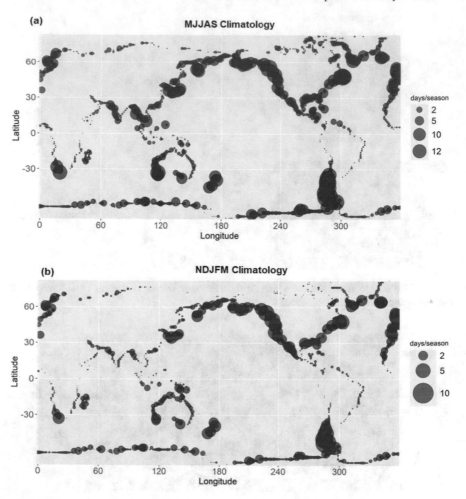

Fig. 2.3 Frequency of landfalling ARs in **a** MJJAS climatology, **b** NDJFM climatology

2.4.1 Global Perspectives

The ARDM proposed by [29] was employed on ERA-Interim IVT data spanning from 1979 to 2015, resulting in the generation of a comprehensive global AR catalog. Their observations revealed an average of 11 ARs globally at a time. Different ARDMs proposed in various studies exhibited slight variations in terms of IVT thresholds and geometric criteria. Global analyses of ARs indicate that landfall locations are predominantly along the west coasts of major continents in mid-latitudes of both hemispheres [30, 33, 64] (Fig. 2.3). Coastal areas characterized by higher AR frequencies also exhibit higher landfall frequencies, emphasizing regions where AR impacts are expected to be substantial.

2.4.2 Regional Perspectives

(a) North American West Coast

Research on ARs over North American continent has been a focal point, yielding valuable insights into their climatology and impacts. Neiman et al. [52] pioneered the establishment of AR climatology for the region, defining ARs as filamentary regions of IWV with specific criteria. Subsequent studies affirmed the seasonality and frequency of ARs along coast. Gershunov et al. [20] employed reanalysis data, identifying landfalling ARs based on criteria involving IVT and IWV. Their findings highlighted the seasonal progression of ARs, peaking in frequency in September and October in British Columbia and shifting southward over subsequent months. Variations in the seasonality of AR landfalls emerged due to different detection methodologies, with IWV-based methods identifying more ARs in the warm season and IVT-based methods identifying more ARs in the cool season.

(b) Western United States

Inland-penetrating ARs have defied conventional belief that ARs are confined to coastal regions, notably in western United States. These ARs can exert significant impacts on areas far from the coast, exemplified by the January 2010 AR event that brought extreme weather conditions and inflicted damage in Arizona [51]. The event underscored the significance of specific positioning and topographical factors in AR impacts, including vertical phasing with upper level dynamics, weak static stability, strong water vapor fluxes, and large-scale dynamics. Studies on AR climatology in the western US, utilizing Integrated Vapor Transport (IVT)-based thresholds, revealed a higher frequency of ARs along the coasts of Oregon, Washington, and northern California (Fig. 2.3) [4, 46, 75].

(c) Southeastern United States

Intense rainfall occurrences in the southeastern United States are impacted by different atmospheric phenomena, including cyclones, and mesoscale convective systems (MCSs) [6]. Recent research has shed light on the role of ARs or AR-related features in certain heavy precipitation scenarios, particularly during winter and transition seasons. A study by [44] revealed that approximately 41% of ARs were associated with extreme rainfall events in the southeastern United States.

(d) Europe

The precipitation trends observed in Europe are impacted by the movement of moisture from both the North Atlantic Ocean and the Mediterranean Sea [17, 33]. ARs play an important role in Europe's hydrological cycle. During winter, the moisture originating from the North Atlantic Ocean significantly impacts precipitation patterns across the entire European continent, whereas the Mediterranean Sea serves as a moisture source that influences southern Europe throughout the year [11, 69]. Unlike North America, ARs in Europe are typically linked with extratropical cyclones and

have received less attention. Nevertheless, research has demonstrated their significance in extreme precipitation events, such as a study in Norway connecting an AR
to heavy rainfall in [48].

The North Atlantic Basin has given specific attention to study AR climatology
and its consequences for Western Europe in recent research [11]. Moisture origins
for ARs arriving in Western Europe primarily stem from subtropical regions, with
additional input from tropical and mid-latitude areas. It is theorized that latent heat
release within ARs influences the intensification of extratropical cyclones, leading to
impactful weather events in Europe [40]. ARs notably impact European precipitation,
especially during the active storm track in winter. However, their impact lessens
during the summer months as convective weather systems become more prevalent.
Nevertheless, regions such as the Iberian Peninsula, France, the UK, and western
Scandinavia continue to experience significant effects from ARs in Europe [55, 70].

(e) **Southern South America**

The impact of heavy rainfall linked with ARs has garnered less attention in South
America in comparison to North America and Europe. Nonetheless, parallels can be
drawn between the impacts of ARs on the mid-latitude west coasts of both continents,
characterized by extensive mountain ranges on the direct way of ARs. These topographical features facilitate orographic precipitation processes, leading to intensified
rainfall and snow events [86]. However, specific local variations in the topography of
mountain ranges in South and North America, including differences in altitude, orientation, width, and the presence of smaller coastal ranges, result in varied orographic
modulation and precipitation patterns linked to landfalling ARs.

In South America, the west coast encounters substantial wintertime precipitation
events, primarily linked to landfalling ARs, which significantly contribute to the
annual total rainfall and snow accumulation in the Andes (specifically between 30°S
and 37°S). Typically, four to five intense wintertime heavy rainfall events contribute
to approximately 70% of the annual precipitation, with a notable portion of these
events attributed to ARs [1]. The presence of stronger winds and deeper cyclones
associated with landfalling ARs tends to result in higher accumulated precipitation,
and extreme events induced by ARs can lead to floods, landslides, and significant
damages, particularly in regions like central Chile.

(f) **ARs in Polar Regions**

ARs govern an important role in the poles by transporting warm, moist air toward
the poles. When ARs reach the ice sheets of Greenland or Antarctica, they result
in orographic uplift due to the steep slopes of these regions [22, 85]. The elevated
terrain of Greenland and East Antarctica amplifies the impact of landfalling ARs on
polar precipitation, akin to their influence on mountainous regions like California or
South America. Along with triggering unusual rainfall, ARs also result in warm-air
advection, giving rise to significant events like the melt event in central Greenland
in July 2012 [85].

2.5 Impacts of Atmospheric Rivers

ARs exert substantial impacts on the regions they traverse, ranging from advantageous to perilous outcomes. These effects are pivotal in the context of flooding and water resource management. ARs frequently interact with mountainous terrain and WCBs in extratropical cyclones, leading to prolonged and intense precipitation [5, 60]. The precipitation they bring, particularly in conjunction with mountainous areas, can result in orographic precipitation enhancement. This process involves the uplift of moist air by elevated terrain, enhancing precipitation efficiency. Various factors, including the stability of the approaching air, wind speed, and characteristics of the underlying terrain, influence the interaction between ARs and mountains. Near-neutral stability often causes the air to flow over the mountain barriers. However, when ARs confront taller mountain ranges like the Sierra Nevada, the formation of the Sierra Barrier Jet is common which results in heavy precipitation [8, 28, 38].

The impact of ARs extends beyond floods, encompassing various aspects of water resources. ARs play a significant role in snowpacks, landslides, groundwater recharge, water quality, ecosystems, and vegetation productivity [5, 60]. The heavy precipitation triggered by ARs has implications for these environmental components. Previous studies have highlighted connections between ARs and damaging winds, suggesting potential effects beyond precipitation extremes.

2.6 Conclusions

The comprehensive exploration of ARs in this article unveils their intricate nature and pivotal role in shaping climate extremes across the globe. By leveraging various observational techniques, including advanced satellite-based observations and reanalysis data, we have significantly enhanced our ability to identify and study ARs on both global and regional scales. This technological progress has opened new avenues for unraveling the complexities inherent in the behavior of ARs and their impacts on climate systems.

Examining ARs from diverse perspectives, ranging from the North American West Coast to Europe, and even polar regions, has underscored their adaptability and distinct influence on regional climate patterns. The intricate interplay of ARs with large-scale climate oscillations further emphasizes the nuanced nature of these atmospheric phenomena. The recognition of ARs as potent drivers of climate extremes is a key takeaway from our exploration. From triggering heavy precipitation and flooding to inducing drought conditions, ARs wield significant influence on the climatic conditions of the regions they impact. As we gain a deeper understanding of these dynamics, the information becomes not only crucial for refining climate models but also indispensable for fortifying early warning systems and developing effective strategies to mitigate the adverse impacts associated with extreme weather events.

In essence, the chapter advocates for the continued pursuit of knowledge surrounding ARs in the broader context of climate extremes. Our evolving comprehension of ARs contributes not only to the scientific understanding of Earth's atmospheric processes but also holds immense practical significance for developing proactive measures to address the challenges posed by extreme weather events in an increasingly variable climate. As we navigate the complexities of ARs, we strengthen our capacity for resilience and informed decision-making, fostering a more sustainable and adaptive approach to global climate challenges.

References

1. R.V. Andreoli, M.T. Kayano, W.H.T. Rego, J. Antunes, The role of the Atlantic multidecadal oscillation precondition in the teleconnection of different El Niño-Southern oscillation types and impacts on the 15 N–15 S South American sector precipitation. 1943–1964 (2020). https://doi.org/10.1002/joc.6309
2. R.V. Blamey, A.M. Ramos, R.M. Trigo, R. Tomé, C.J.C. Reason, The influence of atmospheric rivers over the south atlantic on winter rainfall in South Africa. J. Hydrometeorol. 19(1), 127–142 (2018). https://doi.org/10.1175/JHM-D-17-0111.1
3. F. Cannon, F.M. Ralph, A.M. Wilson, D.P. Lettenmaier, GPM satellite radar measurements of precipitation and freezing level in atmospheric rivers: comparison with ground-based radars and reanalyses. J. Geophys. Res.: Atmos 122(23) (2017). https://doi.org/10.1002/2017JD027355
4. T.W. Corringham, J. McCarthy, T. Shulgina, A. Gershunov, D.R. Cayan, F.M. Ralph, Climate change contributions to future atmospheric river flood damages in the western United States. Sci. Rep. 12(1), 13747 (2022). https://doi.org/10.1038/s41598-022-15474-2
5. H.F. Dacre, P.A. Clark, O. Martinez-Alvarado, M.A. Stringer, D.A. Lavers, How do atmospheric rivers form? Bull. Am. Meteor. Soc. 96(8), 1243–1255 (2015). https://doi.org/10.1175/BAMS-D-14-00031.1
6. N. Debbage, P. Millar, S. Poore, K. Morano, T. Mote, J.M. Sheperd, A climatology of atmospheric river interactions with the southeastern United States coastline. Int. J. Climatol. 37, 4077–4091 (2017). https://doi.org/10.1002/joc.5000
7. D.P. Dee, S.M. Uppala, A.J. Simmons, P. Berrisford, P. Poli, S. Kobayashi, U. Andrae, M.A. Balmaseda, G. Balsamo, P. Bauer, P. Bechtold, A.C.M. Beljaars, L. van de Berg, J. Bidlot, N. Bormann, C. Delsol, R. Dragani, M. Fuentes, A.J. Geer, L. Haimberger, S.B. Healy, H. Hersbach, E.V. Hólm, L. Isaksen, P. Kållberg, M. Köhler, M. Matricardi, A.P. Mcnally, B.M. Monge-Sanz, J.J. Morcrette, B.K. Park, C. Peubey, P. de Rosnay, C. Tavolato, J.N. Thépaut, F. Vitart, The ERA-Interim reanalysis: Configuration and performance of the data assimilation system. Q. J. R. Meteorol. Soc. 137(656), 553–597 (2011). https://doi.org/10.1002/qj.828
8. M.D. Dettinger, Atmospheric rivers as drought busters on the U.S. West Coast. J. Hydrometeorol. 14(6), 1721–1732 (2013). https://doi.org/10.1175/JHM-D-13-02.1
9. M.D. Dettinger, F.M. Ralph, J.J. Rutz, Empirical return periods of the most intense vapor transports during historical atmospheric river landfalls on the U.S. West Coast. J. Hydrometeorol. 19(8), 1363–1377 (2018). https://doi.org/10.1175/JHM-D-17-0247.1
10. D. Dhana Lakshmi, A.N.V. Satyanarayana, Influence of atmospheric rivers in the occurrence of devastating flood associated with extreme precipitation events over Chennai using different reanalysis data sets. Atmos. Res. 215, 12–36 (2019). https://doi.org/10.1016/j.atmosres.2018.08.016
11. B. teau, M. Dournaux, N. Montoux, J.-L. Baray J-L, Atmospheric rivers and associated precipitation over France and Western Europe: 1980–2020. Climatology and case study. Atmosphere (2021)

12. S. Dubey, H. Gupta, M.K. Goyal, N. Joshi, Evaluation of precipitation datasets available on Google earth engine over India. Int. J. Climatol. **41**(10), 4844–4863 (2021). https://doi.org/10. 1002/joc.7102
13. S. Dubey, M.K. Goyal, Glacial lake outburst flood hazard, downstream impact, and risk over the Indian Himalayas. Water Resources Res. **56**(4) (2020). https://doi.org/10.1029/2019WR 026533
14. S. Dubey, A. Sattar, M.K. Goyal, S. Allen, H. Frey, U.K. Haritashya, C. Huggel, Mass movement hazard and exposure in the Himalaya. Earth's Future **11**(9) (2023). https://doi.org/10.1029/202 2EF003253
15. J. Eiras-Barca, N. Lorenzo, J. Taboada, A. Robles, G. Miguez-Macho, On the relationship between atmospheric rivers, weather types and floods in Galicia (NW Spain). Nat. Hazards Earth Syst. Sci. **18**, 1633–1645 (2018). https://doi.org/10.5194/nhess-18-1633-2018
16. R.E.E.N. Ewell, A proposed algorithm for moisture fluxes from atmospheric rivers. Mon. Weather Rev. **3**, 725–735 (1998)
17. Y. Gao, J. Lu, L.R. Leung, Uncertainties in projecting future changes in atmospheric rivers and their impacts on heavy precipitation over Europe. J. Clim. **29**(18), 6711–6726 (2016). https:// doi.org/10.1175/JCLI-D-16-0088.1
18. Y. Gao, J. Lu, L.R. Leung, Q. Yang, S. Hagos, Y. Qian, Dynamical and thermodynamical modulations on future changes of landfalling atmospheric rivers over western North America. Geophys. Res. Lett. **42**(17), 7179–7186 (2015). https://doi.org/10.1002/2015GL065435
19. D. Garaboa-Paz, J. Eiras-Barca, F. Huhn, V. Pérez-Muñuzuri, Lagrangian coherent structures along atmospheric rivers. Chaos (Woodbury, N.Y.) **25**(6), 63105 (2015). https://doi.org/10. 1063/1.4919768
20. A. Gershunov, T. Shulgina, F.M. Ralph, D.A. Lavers, J.J. Rutz, Assessing the climate-scale variability of atmospheric rivers affecting western North America. Geophys. Res. Lett. **44**, 7900–7908 (2017). https://doi.org/10.1002/2017GL074175
21. L. Gimeno, R. Nieto, M. Vázquez, D.A. Lavers (2014). Atmospheric rivers: A mini-review. Frontiers in Earth Science, 2(March), 1–6. https://doi.org/10.3389/feart.2014.00002
22. I.V. Gorodetskaya, M. Tsukernik, K. Claes, M.F. Ralph, W.D. Neff, N.P.M. Van Lipzig, The role of atmospheric rivers in anomalous snow accumulation in East Antarctica. Geophys. Res. Lett. **41**(17), 6199–6206 (2014). https://doi.org/10.1002/2014GL060881
23. M.K. Goyal, C.S.P. Ojha, Evaluation of various linear regression methods for downscaling of mean monthly precipitation in arid Pichola Watershed. Nat. Resourc. **01**(01), 11–18 (2010). https://doi.org/10.4236/nr.2010.11002
24. M.K. Goyal, V. Poonia, V. Jain, Three decadal urban drought variability risk assessment for Indian smart cities. J. Hydrol. **625**, 130056 (2023). https://doi.org/10.1016/j.jhydrol.2023. 130056
25. B. Guan, D.E. Waliser, Atmospheric rivers in 20 year weather and climate simulations: a multimodel, global evaluation. J Geophys. Res. Atmos. **122**(11), 5556–5581 (2017). https:// doi.org/10.1002/2016JD026174
26. B. Guan, Tracking atmospheric rivers globally : spatial distributions and temporal evolution of life cycle characteristics. J. Geophys. Res. Atmos. (2020). https://doi.org/10.1029/2019JD 031205
27. B. Guan, N. Molotch, D. Waliser, F. Fetzer, P. Neiman, Extreme snowfall events linked to atmospheric rivers and surface air temperature via satellite measurements. Geophys. Res. Lett. **37**. https://doi.org/10.1029/2010GL044696
28. B. Guan, N.P. Molotch, D.F. Waliser, E.J. Fetzer, P.J. Neiman, The 2010/2011 snow season in California's Sierra Nevada: role of atmospheric rivers and modes of large-scale variability. Water Resourc. Res. **49**(10), 6731–6743 (2013). https://doi.org/10.1002/wrcr.20537
29. B. Guan, D.F. Waliser, Detection of atmospheric rivers: evaluation and application of an algorithm for global studies. J. Geophys. Res. Atmos. **120**(24), 12514–12535 (2015). https://doi. org/10.1002/2015JD024257
30. B. Guan, D.E. Waliser, Tracking atmospheric rivers globally: spatial distributions and temporal evolution of life cycle characteristics. J. Geophys. Res. Atmos. **124**, 12 523–12 552 (2019). https://doi.org/10.1029/2019JD031205

31. K. Guirguis, A. Gershunov, T. Shulgina, R.E.S. Clemesha, F.M. Ralph, Atmospheric rivers impacting Northern California and their modulation by a variable climate. Clim. Dyn. **52**(11), 6569–6583 (2019). https://doi.org/10.1007/s00382-018-4532-5

32. K. Guirguis, A. Gershunov, R.E.S. Clemesha, T. Shulgina, A.C. Subramanian, F.M. Ralph, Circulation drivers of atmospheric rivers at the North American West Coast. Geophys. Res. Lett. **45**, 12 576–12 584 (2018). https://doi.org/10.1029/2018GL079249

33. M. Ionita, V. Nagavciuc, B. Guan, Rivers in the sky, flooding on the ground: the role of atmospheric rivers in inland flooding in central Europe. Hydrol. Earth Syst. Sci. **24**(11), 5125–5147 (2020). https://doi.org/10.5194/hess-24-5125-2020

34. S. Jha, J. Das, A. Sharma, B. Hazra, M.K. Goyal, Probabilistic evaluation of vegetation drought likelihood and its implications to resilience across India. Glob. Planet. Change **176**, 23–35 (2019). https://doi.org/10.1016/j.gloplacha.2019.01.014

35. S. Jha, J. Das, M.K. Goyal, Low frequency global-scale modes and its influence on rainfall extremes over India: nonstationary and uncertainty analysis. Int. J. Climatol. **41**(3), 1873–1888 (2021). https://doi.org/10.1002/joc.6935

36. Y. Kamae, W. Mei, S.-P. Xie, Ocean warming pattern effects on future changes in East Asian atmospheric rivers. Environ. Res. Lett. **14**(5), 054019 (2019). https://doi.org/10.1088/1748-9326/ab128a

37. S.Y. Kim, A. Upneja, Majority voting ensemble with a decision trees for business failure prediction during economic downturns. J. Innov. Knowl. **6**(2), 112–123 (2021). https://doi.org/10.1016/j.jik.2021.01.001

38. D.E. Kingsmill, P.J. Neiman, B.J. Moore, M. Hughes, S.E. Yuter, F.M. Ralph, Kinematic and thermodynamic structures of sierra barrier jets and overrunning atmospheric rivers during a landfalling winter storm in Northern California. Mon. Weather Rev. **141**(6), 2015–2036 (2013). https://doi.org/10.1175/MWR-D-12-00277.1

39. N. Kumar, P. Patel, S. Singh, M.K. Goyal, Understanding non-stationarity of hydroclimatic extremes and resilience in Peninsular catchments, India. Sci. Rep **13**(1), 12524 (2023). https://doi.org/10.1038/s41598-023-38771-w

40. D.A. Lavers, G. Villarini, The contribution of atmospheric rivers to precipitation in Europe and the United States. J. Hydrol. **522**, 382–390 (2015). https://doi.org/10.1016/j.jhydrol.2014.12.010

41. D.A. Lavers, G. Villarini, R.P. Allan, E.F. Wood, A.J. Wade, The detection of atmospheric rivers in atmospheric reanalyses and their links to British winter floods and the large-scale climatic circulation. J. Geophys. Res. Atmos. **117**(D20) (2012). https://doi.org/10.1029/2012JD018027

42. J. Liang, Y. Yong, Climatology of atmospheric rivers in the Asian monsoon region. Int. J. Climatol. **41**(S1), E801–E818 (2021). https://doi.org/10.1002/joc.6729

43. J. Liang, Y. Yong, M.K. Hawcroft, Long-term trends in atmospheric rivers over East Asia. Clim. Dyn. **60**(3–4), 643–666 (2023). https://doi.org/10.1007/s00382-022-06339-5

44. K. Mahoney, Understanding the role of atmospheric rivers in heavy precipitation in the southeast United States. Mon. Wea. Rev. **144**, 1617–1632 (2016). https://doi.org/10.1175/MWR-D-15-0279.1

45. S.S. Mahto, M.A. Nayak, D.P. Lettenmaier, V. Mishra, Atmospheric rivers that make landfall in India are associated with flooding. Commun. Earth Environ. **4**(1), 120 (2023). https://doi.org/10.1038/s43247-023-00775-9

46. G.J. McCabe, M.D. Dettinger, Decadal variations in the strength of ENSO teleconnections with precipitation in the western United States. Int. J. Climatol. **19**(13), 1399–1410 (1999). https://doi.org/10.1002/(SICI)1097-0088(19991115)19:13<1399::AID-JOC457>3.0.CO;2-A

47. S. Meghani, S. Singh, N. Kumar, M.K. Goyal, Predicting the spatiotemporal characteristics of atmospheric rivers: a novel data-driven approach. Glob. Planet. Change **231**, 104295 (2023). https://doi.org/10.1016/j.gloplacha.2023.104295

48. C. Michel, A. Sorteberg, S. Eckhardt, C. Weijenborg, A. Stohl, M. Cassiani, Characterization of the atmospheric environment during extreme precipitation events associated with atmospheric rivers in Norway—seasonal and regional aspects. Weather Clim. Extremes **34**, 100370 (2021). https://doi.org/10.1016/j.wace.2021.100370

49. B.D. Mundhenk, E.A. Barnes, E.D. Maloney, All-season climatology and variability of atmospheric river frequencies over the North Pacific. J. Clim. **29**, 4885–4903 (2016). https://doi.org/10.1175/JCLI-D-15-0655.1

50. D. Nash, D. Waliser, B. Guan, H. Ye, F.M. Ralph, The role of atmospheric rivers in extratropical and polar hydroclimatology. J. Geophys. Res. Atmos. **123**, 6804–6821 (2018). https://doi.org/10.1029/2017JD028130

51. P.J. Neiman, M. Hughes, B.J. Moore, F.M. Ralph, E.M. Sukovich, Sierra barrier jets, atmospheric rivers, and precipitation characteristics in Northern California: a composite perspective based on a network of wind profilers. Mon. Weather Rev. **141**(12), 4211–4233 (2013). https://doi.org/10.1175/MWR-D-13-00112.1

52. P.J. Neiman, F.M. Ralph, G.A. Wick, J.D. Lundquist, M.D. Dettinger, Meteorological characteristics and overland precipitation impacts of atmospheric rivers affecting the west coast of North America based on eight years of SSM/I satellite observations. J. Hydrometeorol. Am. Meteorol. Soc. **9**(1), 22–47 (2008). https://doi.org/10.1175/2007JHM855.1

53. A.E. Payne, M.-E. Demory, L.R. Leung, A.M. Ramos, C.A. Shields, J.J. Rutz, N. Siler, G. Villarini, A. Hall, F.M. Ralph, Responses and impacts of atmospheric rivers to climate change. Nat. Rev. Earth Environ. **1**(3), 143–157 (2020). https://doi.org/10.1038/s43017-020-0030-5

54. A.E. Payne, G. Magnusdottir, Dynamics of landfalling atmospheric rivers over the North Pacific in 30 years of MERRA reanalysis. J. Clim. **27**(18), 7133–7150 (2014). https://doi.org/10.1175/JCLI-D-14-00034.1

55. J.M. Pereira, M. Basto, A.F. da Silva, The logistic lasso and ridge regression in predicting corporate failure. Procedia Econ. Finance **39**, 634–641 (2016). https://doi.org/10.1016/S2212-5671(16)30310-0

56. S.D. Polade, A. Gershunov, D.R. Cayan, M.D. Dettinger, D.W. Pierce, Natural climate variability and teleconnections to precipitation over the Pacific-North American region in CMIP3 and CMIP5 models. Geophys. Res. Lett. **40**(10), 2296–2301 (2013). https://doi.org/10.1002/grl.50491

57. V. Poonia, M. Kumar Goyal, S. Jha, S. Dubey, Terrestrial ecosystem response to flash droughts over India. J. Hydrol. **605**, 127402 (2022). https://doi.org/10.1016/j.jhydrol.2021.127402

58. H.D. Prince, N.J. Cullen, P.B. Gibson, J. Conway, D.G. Kingston, A Climatology of atmospheric rivers in New Zealand. J. Clim. **34**(11), 4383–4402 (2021). https://doi.org/10.1175/JCLI-D-20-0664.1

59. S. Rakkasagi, M.K. Goyal, S. Jha, Evaluating the future risk of coastal Ramsar wetlands in India to extreme rainfalls using fuzzy logic. J. Hydrol. **632**, 130869 (2024). https://doi.org/10.1016/j.jhydrol.2024.130869

60. F.M. Ralph, M.D. Dettinger, Storms, floods, and the science of atmospheric rivers. Eos, Trans. Am. Geophys. Union **92**(32), 265–266 (2011). https://doi.org/10.1029/2011EO320001

61. F.M. Ralph, P.J. Neiman, R. Rotunno, Dropsonde observations in low-level jets over the northeastern pacific ocean from CALJET-1998 and PACJET-2001: mean vertical-profile and atmospheric-river characteristics. Mon. Weather Rev. **133**(4), 889–910 (2005). https://doi.org/10.1175/MWR2896.1

62. F.M. Ralph, P.J. Neiman, G.A. Wick, S.I. Gutman, M.D. Dettinger, D.R. Cayan, A.B. White, Flooding on California's russian river: role of atmospheric rivers. Geophys. Res. Lett. **33**(13) (2006). https://doi.org/10.1029/2006GL026689

63. F.M. Ralph, M.D. Dettinger, M.M. Cairns, T.J. Galarneau, J. Eylander, Defining "Atmospheric River": how the glossary of meteorology helped resolve a debate. Bull. Am. Meteor. Soc. **99**(4), 837–839 (2018). https://doi.org/10.1175/BAMS-D-17-0157.1

64. F.M. Ralph, M. Dettinger, D. Lavers, I.V. Gorodetskaya, A. Martin, M. Viale, A.B. White, N. Oakley, J. Rutz, J.R. Spackman, H. Wernli, J. Cordeira, Atmospheric Rivers emerge as a global science and applications focus. Bull. Am. Meteor. Soc. **98**(9), 1969–1973 (2017). https://doi.org/10.1175/BAMS-D-16-0262.1

65. F.M. Ralph, M.D. Dettinger, J.J. Rutz, D.E. Waliser (eds.), *Atmospheric Rivers* (Springer International Publishing, Cham, 2020)

66. F.M. Ralph, P.J. Neiman, G.N. Kiladis, K. Weickmann, D.W. Reynolds, A multiscale observational case study of a pacific atmospheric river exhibiting tropical-extratropical connections and a mesoscale frontal wave. Mon. Weather Rev. **139**(4), 1169–1189 (2011). https://doi.org/10.1175/2010MWR3596.1

67. F.M. Ralph, P.J. Neiman, G.A. Wick, Satellite and CALJET aircraft observations of atmospheric rivers over the eastern north pacific ocean during the winter of 1997/98. Mon. Weather Rev. **132**(7), 1721–1745 (2004). https://doi.org/10.1175/1520-0493(2004)132%3c1721:SACAOO%3e2.0.CO;2

68. F.M. Ralph, A.M. Wilson, T. Shulgina, B. Kawzenuk, S. Sellars, J.J. Rutz, M.A. Lamjiri, E.A. Barnes, A. Gershunov, B. Guan, K.M. Nardi, T. Osborne, G.A. Wick, ARTMIP-early start comparison of atmospheric river detection tools: how many atmospheric rivers hit northern California's Russian River watershed? Clim. Dyn. **52**(7–8), 4973–4994 (2019). https://doi.org/10.1007/s00382-018-4427-5

69. A.M. Ramos, R. Tomé, R.M. Trigo, M.L.R. Liberato, J.G. Pinto, Projected changes in atmospheric rivers affecting Europe in CMIP5 models. Geophys. Res. Lett. **43**(17), 9315–9323 (2016). https://doi.org/10.1002/2016GL070634

70. A.M. Ramos, R.M. Trigo, M.L.R. Liberato, R. Tomé, Daily precipitation extreme events in the Iberian Peninsula and its association with atmospheric rivers. J. Hydrometeorol. **16**(2), 579–597 (2015). https://doi.org/10.1175/JHM-D-14-0103.1

71. A.M. Ramos, R.C. Blamey, I. Algarra, R. Nieto, L. Gimeno, R. Tomé, C.J.C. Reason, R.M. Trigo, From Amazonia to southern Africa: atmospheric moisture transport through low-level jets and atmospheric rivers. Ann. N. Y. Acad. Sci. **1436**, 217–230 (2019). https://doi.org/10.1111/nyas.13960

72. K.S. Rautela, J.C. Kuniyal, M.K. Goyal, N. Kanwar, A.S. Bhoj, Assessment and modelling of hydro-sedimentological flows of the eastern river Dhauliganga, north-western Himalaya, India. Nat. Hazards (2024). https://doi.org/10.1007/s11069-024-06413-7

73. K.S. Rautela, S. Singh, M.K. Goyal, Characterizing the spatio-temporal distribution, detection, and prediction of aerosol atmospheric rivers on a global scale. J. Environ. Manage. **351**, 119675 (2024). https://doi.org/10.1016/j.jenvman.2023.119675

74. K.J. Reid, A.D. King, T.P. Lane, D. Hudson, Tropical, subtropical, and extratropical atmospheric rivers in the Australian Region. J. Clim. **35**(9), 2697–2708 (2022). https://doi.org/10.1175/JCLI-D-21-0606.1

75. J.J. Rutz, W.J. Steenburgh, Quantifying the role of atmospheric rivers in the interior western United States. Atmos. Sci. Lett. **13**(4), 257–261 (2012). https://doi.org/10.1002/asl.392

76. J.J. Rutz, W.J. Steenburgh, F.M. Ralph, The inland penetration of atmospheric rivers over Western North America: a lagrangian analysis. Mon. Weather Rev. **143**(5), 1924–1944 (2015). https://doi.org/10.1175/MWR-D-14-00288.1

77. A. Sharma, M.K. Goyal, Assessment of drought trend and variability in India using wavelet transform. Hydrol. Sci. J. **65**(9), 1539–1554 (2020). https://doi.org/10.1080/02626667.2020.1754422

78. C.A. Shields, J.J. Rutz, L.R. Leung, F.M. Ralph, M. Wehner, T. O'Brien, Defining uncertainties through comparison of atmospheric river tracking methods. Bull. Am. Meteorol. Soc **100**, ES93–ES96 (2019). https://doi.org/10.1175/BAMS-D-18-0200.1

79. J. Shu, A.Y. Shamseldin, E. Weller, The impact of atmospheric rivers on rainfall in New Zealand. Sci. Rep. **11**(1), 5869 (2021). https://doi.org/10.1038/s41598-021-85297-0

80. S. Singh, M.K. Goyal, Enhancing climate resilience in businesses: the role of artificial intelligence. J. Clean. Prod. 138228 (2023)

81. S. Singh, N. Kumar, M.K. Goyal, S. Jha, Relative influence of ENSO, IOD, and AMO over spatiotemporal variability of hydroclimatic extremes in Narmada basin, India. AQUA—Water Infrastruct. Ecosyst. Soc.**72**(4), 520–539 (2023). https://doi.org/10.2166/aqua.2023.219

82. S. Singh, M.K. Goyal, An innovative approach to predict atmospheric rivers: exploring convolutional autoencoder. Atmos. Res. **289**, 106754 (2023)

83. S. Singh, M.K. Goyal, S. Jha, Role of large-scale climate oscillations in precipitation extremes associated with atmospheric rivers: nonstationary framework. Hydrol. Sci. J. **68**(3), 395–411 (2023)

84. H. Sodemann, A. Stohl, Moisture origin and meridional transport in atmospheric rivers and their association with multiple cyclones. Mon. Weather Rev. **141**(8), 2850–2868 (2013). https://doi.org/10.1175/MWR-D-12-00256.1

85. J. Turner, H. Lu, J.C. King, S. Carpentier, M. Lazzara, T. Phillips, J. Wille, An extreme high temperature event in coastal East Antarctica associated with an atmospheric river and record summer downslope winds. Geophys. Res. Lett. **49**(4) (2022). https://doi.org/10.1029/2021GL097108

86. M. Viale, R. Valenzuela, R.D. Garreaud, F.M. Ralph, Impacts of atmospheric rivers on precipitation in southern South America. J. Hydrometeorol. **19**(10), 1671–1687 (2018). https://doi.org/10.1175/JHM-D-18-0006.1

87. G.A. Wick, Y.-H. Kuo, F.M. Ralph, T.-K. Wee, P.J. Neiman (2008). Intercomparison of integrated water vapor retrievals from SSM/I and COSMIC. Geophys. Res. Lett. **35**(21), L21805. https://doi.org/10.1029/2008GL035126

88. G.A. Wick, P.J. Neiman, F.M. Ralph, Description and validation of an automated objective technique for identification and characterization of the integrated water vapor signature of atmospheric rivers. IEEE Trans. Geosci. Remote Sens. **51**, 2166–2176 (2013). https://doi.org/10.1109/TGRS.2012.2211024

89. Y. Zhu, R.E. Newell, Atmospheric rivers and bombs. Geophys. Res. Lett. **21**(18), 1999–2002 (1994). https://doi.org/10.1029/94GL01710

90. Y. Zhu, R.E. Newell, A proposed algorithm for moisture fluxes from atmospheric rivers. Mon. Weather Rev. **126**(3), 725–735 (1998). https://doi.org/10.1175/1520-0493(1998)126%3c0725:APAFMF%3e2.0.CO;2

Chapter 3
Key Characteristics of Atmospheric Rivers and Associated Precipitation

3.1 Introduction

ARs are fascinating and dynamic channels within Earth's atmospheric circulation, acting as conduits of concentrated moisture transport that profoundly influence regional precipitation patterns and weather systems [8, 26, 36]. These atmospheric phenomena play a pivotal role in shaping the hydrological cycle and are intimately connected to a complex web of atmospheric and oceanic processes [22, 23, 25]. The spatial and temporal distribution of AR events is intricately linked to various factors, including seasonal weather variations, oceanic temperature anomalies such as El Niño events, and broader climate change dynamics [9, 15, 20]. Regions highly susceptible to the influence of ARs, such as the West Coast of North America and selected areas in Europe, experience heightened frequencies of these phenomena [2, 5–7, 11, 12, 29]. This heightened frequency is often observed during specific seasons when atmospheric conditions are conducive to the formation and propagation of ARs [4, 18]. The consequential influx of moisture from ARs significantly contributes to the annual precipitation totals in these regions, playing a critical role in replenishing water resources and sustaining ecosystems. However, the intense nature of AR-associated precipitation can also lead to challenges such as flooding, especially in areas with inadequate infrastructure and land use practices. The intensity of AR events, gauged by metrics such as moisture transport rates and precipitation levels, serves as a critical determinant of the severity of associated impacts [28, 33, 38]. When ARs unleash their moisture-laden torrents, regions predisposed to heavy rainfall and flooding become particularly vulnerable, facing a cascade of challenges that strain infrastructure, saturate soils, and exacerbate hydrological risks [15, 17, 22, 23, 27]. The complex interplay of atmospheric dynamics unfolds when intense ARs interact with local terrain, resulting in powerful rainfall that can rapidly alter landscapes within a short span of time. This rapid transformation not only challenges the resilience of natural ecosystems but also tests the adaptive capacities of human settlements, highlighting the urgent need for robust disaster preparedness strategies [1, 3, 13, 14, 24, 32, 37].

A detailed comprehension of how often landfalling ARs occur and their specific attributes is vital for accurately assessing flood risks, managing water resources efficiently, and planning for disasters [19, 34]. As these atmospheric systems make landfall, they can unleash torrents of precipitation that inundate low-lying areas, overwhelm drainage systems, and trigger flash floods. The cumulative impacts of landfalling ARs extend beyond immediate flooding events, affecting soil moisture levels, groundwater recharge rates, and overall hydrological balance [2, 31]. This underscores the importance of integrating meteorological data, hydrological models, and risk assessment frameworks to develop proactive strategies that mitigate the impacts of AR-induced flooding and enhance community resilience.

Noteworthy research contributions by experts such as [10, 16, 21, 23, 25, 30, 35, 39] offer valuable insights into the behavior of ARs and their implications for climate dynamics and hydrological processes. These pioneering studies delve into the intricate interplay between atmospheric moisture transport, precipitation patterns, and regional climate variability, shedding light on the underlying mechanisms driving AR variability and evolution. By leveraging these research findings, scientists and policymakers can craft targeted strategies tailored to local vulnerabilities, bolstering adaptive capacities, and enhancing disaster resilience in the face of evolving climate scenarios. By delving deeply into the intricate nuances of AR behavior and associated precipitation patterns, scientists and policymakers can craft targeted strategies to mitigate the impacts of these atmospheric phenomena. Continued research endeavors focused on ARs are essential for advancing our comprehension of these intricate systems, refining forecasting capabilities for extreme weather events linked to ARs, and bolstering resilience in the face of evolving climate scenarios.

3.2   AR Key Characteristics

3.2.1   AR Frequency

The frequency, duration, and intensity of ARs have been analyzed to study their impacts. AR frequency represents the frequency of AR condition at each location (grid) and has been expressed as a percentage of timesteps.

In other words, the percentage of timesteps, a grid representing AR condition in 6 hourly AR detection for the period of 40 years from 1979 to 2018, is taken as the AR frequency of the grid. Western and eastern North America, Europe, east Asia, Australia, New Zealand, Southernmost South America, and Africa experience a significant amount of ARs in a year. These ARs, if sustained for a longer duration at a region, could bring beneficial to disastrous impacts over these regions.

The regions with dark green color having MK z statistics less than -1.96 represent a significant decreasing trend and red color with "z" greater than 1.96 represent

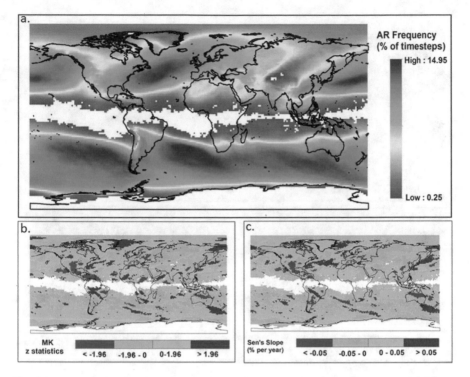

Fig. 3.1 **a** Frequency of ARs at each grid expressed in percent of timesteps; **b** Mann–Kendall trend test's "z" statistics and **c** Sen's slope ("s") at a 5% level of significance

a significant increasing trend in the data, whereas light green and yellow colors represent decreasing and increasing trend but not significant. We observed that there is an overall increasing trend in the AR frequency, especially at regions with higher frequency but not very statistically significant except for a few regions (Fig. 3.1). The spatial distribution of Sen's slope statistics shows approximately a 0.05% of increment in AR frequency at regions with statistically increasing and decreasing trends.

3.2.2 AR Duration

Average AR duration has been calculated as an average time for a grid representing AR for consecutive continuous timesteps throughout the period of AR detection (Fig. 3.2). For example, a grid with 20 h of average AR duration represents that, on average that grid was consistently representing a part of an AR at various timesteps throughout the detection period. This assessment gives us a comparatively better understanding of AR-prone regions. American meteorological society classifies it from weak to exceptional. The west coast of North and South America, Europe,

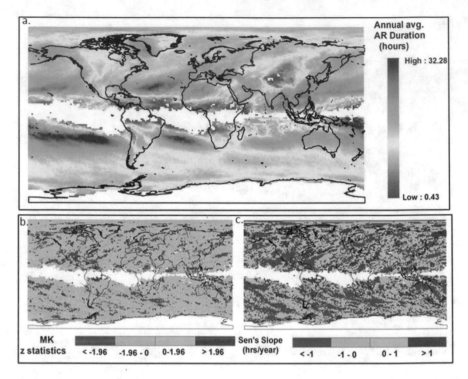

Fig. 3.2 a Duration of ARs in terms of average time sustained over a region; **b** Mann–Kendall trend test's "z" statistics and Sen's slope ("s") at a 5% level of significance

Southeast Asia, and Australia have an average AR duration of 20 to 30 h which may result in beneficial to the devastating type of AR precipitation in the region. AR frequency represents the overall spatial distribution of ARs for the detection period, whereas AR duration represents the average lasting time duration of a grid representing AR which is comparatively a discrete-instantaneous behavior of ARs.

Figure 3.2b (z statistics) shows an overall decreasing trend (not statistically significant) in annual average AR duration, it can be observed here also that the region with high duration distribution has an increasing trend, but only a few regions have a statistically significant increasing trend. Regions with increasing trends have been noted with a rate of 1 h or more per year in AR duration.

3.2.3   AR Intensity

AR intensity is of the main measures of AR moisture-carrying capacity and is represented by the variation in the magnitude of the IVT of collective grids representing an AR. The mean IVT magnitude of grids being identified as an object at any timestep in AR detection is represented as AR intensity. The annual average AR intensity

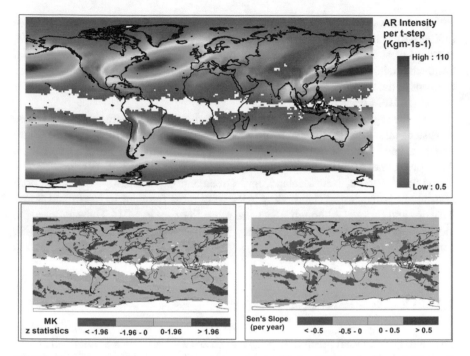

Fig. 3.3 Intensity of ARs represented as an average magnitude of IVT at the time of AR condition over a region; Mann–Kendall trend test's "z" statistics and Sen's slope ("s") at 5% level of significance

per step, represented in Fig. 3.3, indicates the variation in the magnitude of mean IVT at the locations of AR detected. The higher value of intensity represents a high probability of moisture density at that location. The AR shape corresponding to the region of the higher magnitude of intensity is most probable for bringing huge floods at landfalling. The regions with a higher magnitude of AR intensity (west coast of North America and Europe, southmost South America, east coast of Asia, southmost Australia, etc.) coincide with regions of higher AR frequency which means regions with high intensity have more probability of high intensity as well.

The intensity of ARs has been found to increase as of frequency of ARs but only in some regions of high intensity, and it was statistically significant in only some locations of mid-latitude at a 5% level of significance (Fig. 3.3).

3.2.4 Frequency of Landfalling ARs

Since these ARs bring floods and may be responsible for droughts in some regions, it is important to understand the spatiotemporal characteristics of landfalling ARs.

These ARs were noticed almost in all the continents, but the frequency of land-falling ARs was comparatively higher on the west and east coast of North America, southern-west South America, the west coast of Europe, South Asia, East Asia, Southeast Asia, Australia, New Zealand, etc. We observed that these ARs were more frequent in the summer season (MJJAS climatology) in comparison with the winter season (NDJFM climatology) (Fig. 3.4).

Also, the regions with a high frequency of landfalling ARs were found to be more consistent on a decadal scale. The eastern and western coasts of North and South America, southernmost South America, and the west coast of Europe experience ARs for around 11–20 days in both seasons in a year (Fig. 3.4). Whereas South-east Asia, Australia, New Zealand, and Greenland experience 5–10 ARs in both seasons. In both seasons, landfalling ARs were most frequent in North and South America and Europe. Although ARs were more frequent in MJJAS climatology, the most significant increase in AR frequency was observed in the Asian and Australian continents.

3.3   AR-Associated Precipitation

The daily gridded total precipitation dataset from ECMWF (ERA5) obtained at $1.5°$ spatial resolution has been tracked with the availability of AR at each grid for the same day to explore its connectivity with AR. The precipitation received at a location, if found coinciding with the availability of AR condition sustaining for at least one time step (6-h), is considered AR-associated precipitation. For this, the spatial distribution of detected ARs has been superimposed with the total precipitation data obtained from ECMWF at the same timestep, and the fraction of total precipitation as AR-associated precipitation has been estimated (Fig. 3.5).

In the case of NDJFM seasonal climatology, the AR average daily precipitation is concentrated over the west coast of North America and Europe as well as the east coast of South America and Australia (Fig. 3.5c). Whereas, in MJJAS seasonal climatology, AR precipitation is more evident over central-east Asia, New Zealand, southmost of North America, and the southwest region of Australia (Fig. 3.5b) Since in some of the cases, even if an AR was present over a region, it may not necessarily produce precipitation unless some instability in the flow of the moisture (part of an AR) is caused, and if it brings precipitation in certain atmospheric situations, it will not be spatially equally distributed just underneath of the AR elongation.

Therefore, in many cases, even if an AR was consistently available, the precipitation received was null, and it resulted in a lesser magnitude of AR-associated precipitation at a region. While these ARs could bring as much as hundreds of centimeters of precipitation to a location in some specific situations. These plots could give an impression of the regions globally receiving precipitation throughout the globe to further explore the detailed contribution of precipitation in different seasons through these ARs (Fig. 3.5).

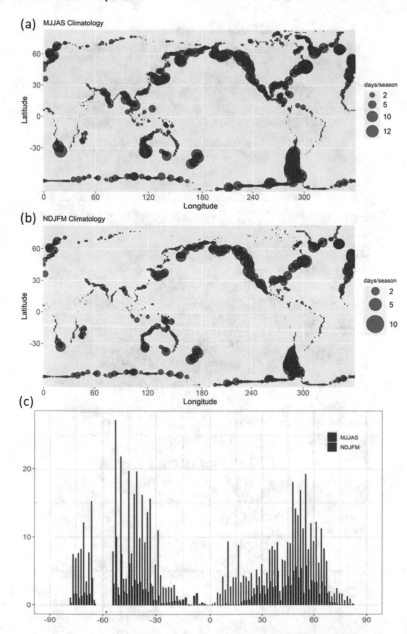

Fig. 3.4 Frequency of landfalling AR intersecting coasts in days per season **a** MJJAS, **b** NDJFM, **c** AR frequency with respect to the latitude of ARs making intersections at coastlines

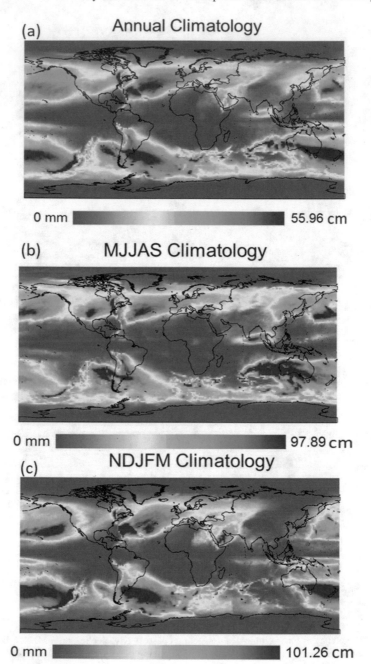

Fig. 3.5 Precipitation associated with ARs for **a** annual climatology, **b** MJJAS climatology, and **c**NDJFM climatology

3.4 Conclusions

ARs are vital components of Earth's atmospheric dynamics, significantly influencing regional precipitation patterns and posing both benefits and challenges to affected regions by producing beneficial rain to disastrous floods. The chapter underscores the importance of understanding the spatial and temporal distribution of AR events, their impacts on hydrological cycles, and their association with atmospheric and oceanic processes. Along with the West Coast of North America and Europe, there are several regions susceptible to AR influence, and experience significant AR frequencies during specific seasons, contributing significantly to annual precipitation totals while also presenting risks like flooding due to intense rainfall.

The intensity of AR events, gauged by moisture transport rates and precipitation levels, plays a pivotal role in determining the severity of associated impacts, particularly in regions prone to heavy rainfall and flooding. Effective flood risk assessment, water resource management, and disaster preparedness planning necessitate a nuanced understanding of AR characteristics and landfall patterns. Integrating meteorological data, hydrological models, and risk assessment frameworks is essential for developing proactive strategies to mitigate AR-induced flooding impacts and enhance community resilience.

Continued research efforts are crucial for advancing our knowledge of AR behavior, refining forecasting capabilities for extreme weather events linked to ARs, and bolstering resilience in the face of changing climate conditions. By delving deeper into the complexities of AR phenomena, we can better prepare for and respond to the impacts of these significant atmospheric events, ultimately enhancing our ability to adapt to a changing climate landscape.

References

1. J. Das, S. Jha, M.K. Goyal, On the relationship of climatic and monsoon teleconnections with monthly precipitation over meteorologically homogenous regions in India: Wavelet & global coherence approaches. Atmos. Res. **238**, 104889 (2020). https://doi.org/10.1016/j.atmosres.2020.104889
2. M.D. Dettinger, Atmospheric rivers as drought busters on the U.S. West Coast. J. Hydrometeorol. **14**(6), 1721–1732 (2013). https://doi.org/10.1175/JHM-D-13-02.1
3. S. Dubey, M.K. Goyal, Glacial lake outburst flood hazard, downstream impact, and risk over the Indian Himalayas. Water Resourc. Res. **56**(4) (2020). https://doi.org/10.1029/2019WR026533
4. V. Espinoza, D.E. Waliser, B. Guan, D.A. Lavers, F.M. Ralph, Global analysis of climate change projection effects on atmospheric rivers. Geophys. Res. Lett. **45**(9), 4299–4308 (2018). https://doi.org/10.1029/2017GL076968
5. Y. Gao, J. Lu, L.R. Leung, Uncertainties in projecting future changes in atmospheric rivers and their impacts on heavy precipitation over Europe. J. Clim. **29**(18), 6711–6726 (2016). https://doi.org/10.1175/JCLI-D-16-0088.1
6. A. Gershunov, T. Shulgina, R.E.S. Clemesha, K. Guirguis, D.W. Pierce, M.D. Dettinger, D.A. Lavers, D.R. Cayan, S.D. Polade, J. Kalansky, F.M. Ralph, Precipitation regime change in Western North America: the role of atmospheric rivers. Sci. Rep. **9**(1), 9944 (2019). https://doi.org/10.1038/s41598-019-46169-w

7. A. Gershunov, T. Shulgina, F.M. Ralph, D.A. Lavers, J.J. Rutz, Assessing the climate-scale variability of atmospheric rivers affecting western North America. Geophys. Res. Lett. **44**, 7900–7908 (2017). https://doi.org/10.1002/2017GL074175

8. L. Gimeno, R. Nieto, M. Vázquez, D.A. Lavers, Atmospheric rivers: a mini-review. Front. Earth Sci. **2**, 1–6 (2014). https://doi.org/10.3389/feart.2014.00002

9. B. Guan, D.E. Waliser, Detection of atmospheric rivers: evaluation and application of an algorithm for global studies. J. Geophys. Res. Atmos. **120**(24), 12514–12535 (2015). https://doi.org/10.1002/2015JD024257

10. B. Guan, D.E. Waliser, Atmospheric rivers in 20 year weather and climate simulations: a multimodel, global evaluation. J. Geophys. Res. Atmos. **122**(11), 5556–5581 (2017). https://doi.org/10.1002/2016JD026174

11. K. Guirguis, A. Gershunov, R.E.S. Clemesha, T. Shulgina, A.C. Subramanian, F.M. Ralph, Circulation drivers of atmospheric rivers at the North American West Coast. Geophys. Res. Lett. **45**, 12576–12584 (2018). https://doi.org/10.1029/2018GL079249

12. M. Ionita, V. Nagavciuc, B. Guan, Rivers in the sky, flooding on the ground: the role of atmospheric rivers in inland flooding in central Europe. Hydrol. Earth Syst. Sci. **24**(11), 5125–5147 (2020). https://doi.org/10.5194/hess-24-5125-2020

13. N. Kumar, P. Patel, S. Singh, M.K. Goyal, Understanding non-stationarity of hydroclimatic extremes and resilience in Peninsular catchments, India. Sci. Rep. **13**(1), 12524 (2023). https://doi.org/10.1038/s41598-023-38771-w

14. N. Kumar, V. Poonia, B.B. Gupta, M.K. Goyal, A novel framework for risk assessment and resilience of critical infrastructure towards climate change. Technol. Forecast. Soc. Chang. **165**, 120532 (2021). https://doi.org/10.1016/j.techfore.2020.120532

15. D.A. Lavers, R.P. Allan, E.F. Wood, G. Villarini, D.J. Brayshaw, A.J. Wade, Winter floods in Britain are connected to atmospheric rivers. Geophys. Res. Lett. **38**(23) (2011). https://doi.org/10.1029/2011GL049783

16. D.A. Lavers, G. Villarini, The contribution of atmospheric rivers to precipitation in Europe and the United States. J. Hydrol. **522**, 382–390 (2015). https://doi.org/10.1016/j.jhydrol.2014.12.010

17. J. Liang, Y. Yong, Climatology of atmospheric rivers in the Asian monsoon region. Int. J. Climatol. **41**(S1), E801–E818 (2021). https://doi.org/10.1002/joc.6729

18. E.C. Massoud, V. Espinoza, B. Guan, D.E. Waliser, Global climate model ensemble approaches for future projections of atmospheric rivers. Earth's Future **7**(10), 1136–1151 (2019). https://doi.org/10.1029/2019EF001249

19. S. Meghani, S. Singh, N. Kumar, M.K. Goyal, Predicting the spatiotemporal characteristics of atmospheric rivers: a novel data-driven approach. Global Planet. Change **231**, 104295 (2023). https://doi.org/10.1016/j.gloplacha.2023.104295

20. D. Nash, D. Waliser, B. Guan, H. Ye, F.M. Ralph, The role of atmospheric rivers in extratropical and polar hydroclimatology. J. Geophys. Res. Atmos. **123**, 6804–6821 (2018). https://doi.org/10.1029/2017JD028130

21. P.J. Neiman, M. Hughes, B.J. Moore, F.M. Ralph, E.M. Sukovich, Sierra barrier jets, atmospheric rivers, and precipitation characteristics in Northern California: a composite perspective based on a network of wind profilers. Mon. Weather Rev. **141**(12), 4211–4233 (2013). https://doi.org/10.1175/MWR-D-13-00112.1

22. H. Paltan, D. Waliser, W.H. Lim, B. Guan, D. Yamazaki, R. Pant, S. Dadson, Global floods and water availability driven by atmospheric rivers. Geophys. Res. Lett. **44**(20), 10387–10395 (2017). https://doi.org/10.1002/2017GL074882

23. A.E. Payne, M.-E. Demory, L.R. Leung, A.M. Ramos, C.A. Shields, J.J. Rutz, N. Siler, G. Villarini, A. Hall, F.M. Ralph, Responses and impacts of atmospheric rivers to climate change. Nat. Rev. Earth Environ. **1**(3), 143–157 (2020). https://doi.org/10.1038/s43017-020-0030-5

24. S. Rakkasagi, M.K. Goyal, S. Jha, Evaluating the future risk of coastal Ramsar wetlands in India to extreme rainfalls using fuzzy logic. J. Hydrol. **632**, 130869 (2024). https://doi.org/10.1016/j.jhydrol.2024.130869

25. F.M. Ralph, M.D. Dettinger, Storms, floods, and the science of atmospheric rivers. Eos, Trans. Am. Geophys. Union **92**(32), 265–266 (2011). https://doi.org/10.1029/2011EO320001
26. F.M. Ralph, M.D. Dettinger, J.J. Rutz, D.E. Waliser (eds.), *Atmospheric Rivers* (Springer International Publishing, Cham, 2020)
27. F.M. Ralph, P.J. Neiman, G.A. Wick, S.I. Gutman, M.D. Dettinger, D.R. Cayan, A.B. White, Flooding on California's Russian river: role of atmospheric rivers. Geophys. Res. Lett. **33**(13) (2006). https://doi.org/10.1029/2006GL026689
28. F.M. Ralph, J.J. Rutz, J.M. Cordeira, M. Dettinger, M. Anderson, D. Reynolds, L.J. Schick, C. Smallcomb, A scale to characterize the strength and impacts of atmospheric rivers. Bull. Am. Meteor. Soc. **100**(2), 269–289 (2019). https://doi.org/10.1175/BAMS-D-18-0023.1
29. A.M. Ramos, R.M. Trigo, R. Tomé, M.L.R. Liberato, Impacts of atmospheric rivers in extreme precipitation on the European Macaronesian Islands. Atmosphere **9**, 325 (2018). https://doi.org/10.3390/atmos9080325
30. A.M. Ramos, A.M. Wilson, M.J. Deflorio, M.D. Warner, E. Barnes, R. Garreaud, I.V. Gorodetskaya, B. Moore, A. Payne, C. Smallcomb, D.A. Lavers, M. Wehner, F. Martin, H. Sodemann, 2018 international atmospheric rivers conference: multi-disciplinary studies and high-impact applications of atmospheric rivers 1–8 (2019). https://doi.org/10.1002/asl.935
31. K.S. Rautela, S. Singh, M.K. Goyal, Characterizing the spatio-temporal distribution, detection, and prediction of aerosol atmospheric rivers on a global scale. J. Environ. Manage. **351**, 119675 (2024). https://doi.org/10.1016/j.jenvman.2023.119675
32. A. Sharma, M.K. Goyal, Assessment of ecosystem resilience to hydroclimatic disturbances in India. Global Change Biol. **24**(2) (2018). https://doi.org/10.1111/gcb.13874
33. J. Shu, A.Y. Shamseldin, E. Weller, The impact of atmospheric rivers on rainfall in New Zealand. Sci. Rep. **11**(1), 5869 (2021). https://doi.org/10.1038/s41598-021-85297-0
34. S. Singh, M.K. Goyal, Enhancing climate resilience in businesses: the role of artificial intelligence. J. Cleaner Prod. 138228 (2023)
35. S. Singh, M.K. Goyal, An innovative approach to predict atmospheric rivers: exploring convolutional autoencoder. Atmos. Res. **289**, 106754 (2023)
36. S. Singh, M.K. Goyal, S. Jha, Role of large-scale climate oscillations in precipitation extremes associated with atmospheric rivers: nonstationary framework. Hydrol. Sci. J. **68**(3), 395–411 (2023)
37. S. Singh, N. Kumar, M.K. Goyal, S. Jha, Relative influence of ENSO, IOD, and AMO over spatiotemporal variability of hydroclimatic extremes in Narmada basin, India. AQUA Water Infrastruct. Ecosyst. Soc. (2023). https://doi.org/10.2166/aqua.2023.219
38. M. Viale, R. Valenzuela, R.D. Garreaud, F.M. Ralph, Impacts of atmospheric rivers on precipitation in southern South America. J. Hydrometeorol. Am. Meteorol. Soc. **19**(10), 1671–1687 (2018). https://doi.org/10.1175/JHM-D-18-0006.1
39. Y. Zhu, R.E. Newell, Atmospheric rivers and bombs. Geophys. Res. Lett. **21**(18), 1999–2002 (1994). https://doi.org/10.1029/94GL01710

Chapter 4
Major Large-Scale Climate Oscillations and Their Interactions with Atmospheric Rivers

4.1 Introduction

Climate oscillations play a pivotal role in shaping global weather patterns and influencing various aspects of regional hydrology [6, 28]. These oscillations, characterized by their intricate dynamics, interact with atmospheric rivers (ARs) in ways that have significant implications for climate variability, extreme weather events, and water resource management strategies. Large-scale climate oscillations (LSCOs) are persistent patterns of atmospheric circulation and sea surface temperature (SST) anomalies and can significantly affect weather and climate at different spatial and temporal scales [6, 28]. ARs have been observed to be linked to well-known LSCOs in regions such as North America and Europe, such as El-Nino Southern Oscillation (ENSO), North Atlantic Oscillation (NAO), and Artic Oscillation (AO) [11, 13, 23]. For example, during the winter of 2010–11, the simultaneous negative phase of two LSCOs, the Pacific/North American (PNA) pattern and the AO, was linked to the occurrence of an exceptionally high number of ARs in California's Sierra Nevada [10]. A further study conducted in 2015 by Guan and Waliser showed that ARs became less common in extratropical and subtropical regions during El Niño phases, with notable anomalies found in the North Atlantic and the northeastern Pacific. Furthermore, they observed that negative anomalies in AR frequency follow positive anomalies during phases 5–8 of the Madden Julian Oscillation (MJO), which originate in East Asia and extend south of the Aleutian Islands. Accurately predicting ARs on various time frames requires an understanding of these relationships [21, 27, 29, 30]. To improve our comprehension of the mechanisms influencing AR frequency and precipitation at several regions and climates, more investigation is required.

LSCOs, or atmospheric teleconnection patterns, are long-term patterns of air circulation and sea surface temperature anomalies [26, 28]. They significantly influence weather patterns and climate conditions. Some widely explored LSCOs include ENSO, NAO, PDO, and AO [2, 8, 28, 36]. These LSCOs exhibit characteristic spatial patterns and impact temperature, precipitation, and other climate variables in their

M. K. Goyal and S. Singh, *Understanding Atmospheric Rivers Using Machine Learning*, SpringerBriefs in Applied Sciences and Technology, https://doi.org/10.1007/978-3-031-63478-9_4

respective regions. Studies have shown significant connections between LSCOs and ARs along the west coasts of North America and Europe [11, 13]. Research conducted in Europe has revealed links between LSCOs and ARs, such as the Scandinavian pattern and the NAO [19]. Global analyses of ARs and LSCOs have shown their influences on AR precipitation during elongated season from October to March, which aligns with the peak activity of these modes [13, 25].

During El Niño phases, ARs become less frequent in mid-to-high latitudes, while positive anomalies of AR frequency are associated with specific phases of the Madden–Julian Oscillation (MJO) [11]. The AO and PNA patterns influence AR frequency in various regions, with positive and negative phases leading to different outcomes. Given that the modes operate on distinct timeframes, these relationships between ARs and LSCOs have consequences for sub-seasonal, seasonal, and long-term prediction [9, 12]. The way that AO and PNA affect AR intensity is consistent with how they affect AR precipitation [11, 12]. These modes do, however, have a more localized effect on AR precipitation, indicating the necessity of taking into account both large-scale modes and regional and local-scale variability. To gain a deeper knowledge of the mechanisms determining AR frequency and precipitation across various regions, seasons, and climate regimes, more research is required.

Traditional stationary models assume that the underlying statistical properties of extreme events remain constant over time [16]. However, in many cases, this assumption does not hold due to climate change and other dynamic factors. The stationary limitations arise from the inability to capture the temporal evolution and variability of extremes [4, 38]. Climate change, land use changes, and other anthropogenic factors can significantly alter the distribution and occurrence of extremes [14, 20]. By neglecting nonstationary behaviors, stationary models may provide inaccurate estimations and predictions. The nonstationary analysis addresses this limitation by incorporating physical processes as covariates in GEV models [3, 34]. These covariates can capture the influence of changing climate conditions on extreme events. By considering the temporal evolution of these covariates, nonstationary models can capture the time-varying behavior of extremes and provide more accurate estimations of their magnitudes and frequencies [3, 39].

Modeling nonstationary behavior through GEV has been widely adopted in various fields. Researchers have utilized covariates such as anomalies of climate oscillations (ENSO, PNA, MJO, etc.) and SST to capture the spatiotemporal variability of climate extremes [14, 18, 32, 33]. By incorporating covariates into GEV models, they have successfully demonstrated improvements in estimating extreme event magnitudes and understanding their relationship with changing environmental conditions [24, 37]. For instance, Katz [16] discussed the use of statistics and physical covariates to improve the analysis of extreme climatic applications. They applied the maximum likelihood estimation with covariates such as El-Niño phenomenon to assess extreme precipitation and streamflow. Similarly, Sugahara et al. [35] modeled nonstationary extreme precipitation in Sao Paulo, Brazil, using the General Pareto distribution and found a substantial rise in the degree and occurrence of extremes. Yilmaz and Perera [40] modeled nonstationary through intensity–duration–frequency curves for analyzing extreme rainfall, emphasizing underestimation of extreme precipitation under stationary assumptions. Mondal and Mujumdar [22] modeled extreme precipitation in India, linking physical covariates such as climate oscillation and temperature anomaly to intensity, duration, and frequency. The advantages of nonstationary analysis using GEV lie in its ability to capture dynamics of hydroclimatic extremes and their underlying processes [1, 17]. By considering the temporal variations in covariates, we can account for the changing climate and other factors influencing extreme events. This leads to more accurate assessments of risk and better informed decision-making in various fields, including hydrology, climatology, and engineering.

4.2 Case Study: Relative Role of Climate Oscillations in Precipitation Extremes

4.2.1 Extreme Value Modeling

Extreme value modeling was conducted to assess the impact of LSCOs on precipitation extremes linked with ARs. The choice of nonstationary extreme value modeling was made to account for the time-varying pattern of AR precipitation. This study incorporated four major LSCOs (AO, NAO, ENSO, and PDO) considering them as potential covariates influencing the extreme value distribution of precipitation extremes. The role of LSCOs was examined by comparing the stationary and nonstationary fit in which stationary case was considered when parameters of GEV distribution were comparatively stationary with the time, whereas in nonstationary case, the parameters of GEV distributions were fitting best while incorporating the temporal variability of either LSCO or the combinations of LSCOs in precipitation extremes. Stationarity and nonstationarity induced by these LSCOs over precipitation extremes were assessed using several statistical tests and return levels of these extremes under the discovered influence of these precipitation extremes (PRCPTOT, R95TOT, Rx5day) were examined at 10, 20, 50, and 100 years in two seasons MJJAS and NDJFM.

4.2.2 Nonstationary Extreme Value Modeling

We analyzed probabilistic distribution of precipitation extremes, PRCPTOT, R95TOT, and Rx5day in GEV as continuous random variables. The CDF of the GEV distribution is expressed in Eq. (4.1).

$$F(x, \mu, \sigma, \xi) = \begin{cases} \exp\left\{-\left[1 + \frac{\xi(x-\mu)}{\sigma}\right]^{-1/\xi}\right\}, \sigma > 0, 1 + \frac{\xi(x-\mu)}{\sigma} > 0, \xi \neq 0 \\ \qquad\qquad\qquad \{n \text{ odd} \\ \qquad\qquad \exp\left\{-\exp\left[-\frac{(x-\mu)}{\sigma}\right]\right\}, \sigma > 0, \xi = 0 \end{cases}$$

$$(4.1)$$

In this context, the analysis involves the time sequence of precipitation extremes denoted by x, with μ, σ, and ξ representing the location, scale, and shape parameters of the distribution, respectively. The CDF of x, denoted as $F(x; \mu, \sigma, \xi)$, is determined by these parameters.

In stationarity scenario, the GEV distribution parameters are considered constant, while in nonstationary setup, these parameters vary over time, influenced by either individual or group of multiple LSCOs. However, here nonstationary aspect is applied only to the location and scale parameters, as modeling shape parameter is complex and thus kept as it is to simplify the model. This approach allows for modeling the influence of LSCOs on precipitation extremes using linear combinations of covariates. References to previous works by [7, 14, 15, 17, 18, 33] support this methodology.

$$\text{M2} : X \sim GEV[\mu_0 + \mu_2 C_2, \ \sigma, \ \xi]$$

$$(4.2)$$

$$\text{M20} : X \sim GEV[(\mu_0 + \mu_3 C_3 + \mu_4 C_4), \ (\sigma_0 + \sigma_3 C_3 + \sigma_4 C_4), \ \xi]$$

$$(4.3)$$

Using variables as AO, NAO, ENSO, and PDO as C1, C2, C3, and C4 correspondingly in the models, a total of 28 such linear combinations of covariates as LSCOs have been produced as of Eqs. 4.2 and 4.3 [31].

Here, in Eq. (4.2), the temporal variability of particular precipitation extreme is being governed by the NAO only at that region which is prominently expressed by the change in μ. Similarly, in the 20th combination or model M20, the temporal variability of particular precipitation extreme is being governed by AO and NAO combinedly and it is expressed in terms of variation in the location and scale parameters of GEV distribution of that individual precipitation extreme at that region. In this consideration, 28 such combinations were made in same manner [32].

4.2.3 Estimation of GEV Parameters

The significance of these models was evaluated using likelihood ratio test (LR test) [5]. This test serves to prioritize between stationary and nonstationary cases. The selection criterion is made on the statistic derived from Eq. 4.4:

$$2[nllh(S) - nllh(NS)] > c\alpha$$

$$(4.4)$$

In this instance, *nllh(S)* and *nllh(NS)* denote the negative log-likelihood of stationary and nonstationary cases, respectively, and $c\alpha$ is the $(1-\alpha)$ quantile of the Chi-square distribution. The level of significance (α) was set at 5%. In a given significance level, the difference between the two models was assumed to approximate a chi-squared distribution.

In case nonstationarity is verified, the chi-squared distribution's *p*-value is used to identify the optimal nonstationary model. In case the *p*-value is above 0.05, the stationarity null hypothesis is considered discarded. Every combination of the 28 nonstationary models was subjected to the LR test. Maximum likelihood estimation approach was adopted to estimate parameters. This method is preferred due to its ability to incorporate nonstationarity into the distribution parameters effectively [16].

$$L(\theta) = -n \log - \left(1 + \frac{1}{\xi}\right) \sum_{i=1}^{n} \log\left[1 + \frac{\xi(xi - \mu)}{\sigma}\right]$$

$$- \sum_{i=1}^{n} \log\left[1 + \frac{\xi(xi - \mu)}{\sigma}\right]^{-\left(\frac{1}{\xi}\right)}, 1 + \frac{\xi(xi - \mu)}{\sigma} > 0 \qquad (4.5)$$

The distribution parameters were obtained by minimizing the likelihood function $L(\theta)$, where n denotes size of sample and θ denotes parameter vector.

4.2.4 Relative Influence of Climate Oscillation and Influenced Precipitation Extremes

Using statistical analysis, we compared the GEV fit of precipitation extremes in stationary (assuming GEV parameter as constant over time or not at all influenced by LSCOs) and nonstationary cases (GEV parameters as function of AO, NAO, ENSO, and PDO). Results revealed that PDO (a long-lasting ENSO phenomenon) had the most significant impact on large-scale weather patterns. In about 15% to 30% of areas, variations in extreme precipitation were linked to seasonal PDO indices, with MJJAS indices capturing these variations about 50% better than NDJFM. This suggests that around 30% of the region is strongly influenced by PDO during the summer (MJJAS) [32]. Only a small percentage (less than 1%) of areas showed a good fit with stationary models for instant extreme precipitation (Rx1day and Rx5day), while this increased to about 14% for cumulative extreme precipitation (R95pTOT) across both seasons. Stationary extremes were mostly observed in areas with very little or no historical precipitation data.

Return levels of extreme precipitation were estimated over different time frames to assess vulnerability. Higher magnitudes were observed along the equator and in certain extratropical regions, while polar regions and arid areas showed near-zero return levels due to minimal rain or ice contributions. North America and Europe, parts of South America, Africa, and Australia were identified as more vulnerable to extreme precipitation (Fig. 4.1). The extremes expressed in terms of return levels varied slightly between seasonal covariate combinations, particularly noticeable along equatorial and mid-latitude regions. The high magnitude of extremes associated with MJJAS-LSCOs was mainly observed in South America, East Asia, and Australia, while along equatorial and mid-latitude regions of the southern hemisphere in NDJFM-LSCOs.

4.3   Precipitation Extremes and AR

The precipitation extremes were concentrated around the equator or in specific mid-latitude zones. As ARs play a significant role in transporting moisture poleward and contribute substantially to the total annual water supply in mid-latitudes, we analyzed the variability in spatial extent of these precipitation extremes with the significant availability of ARs.

4.3.1   Precipitation Extremes and AR Duration

The analysis delved into the interplay between extreme precipitation occurrences and ARs, focusing on how AR frequency and duration relate to significant extreme events. By setting thresholds for extreme precipitation levels (Rx1day > 20 mm, Rx5day > 30 mm, R95pTOT > 200 mm), study explored where and when ARs played a pivotal role in generating extreme precipitation. The findings highlighted that longer AR durations tended to align with extreme precipitation events, particularly in certain mid-latitude regions globally. However, it was noted that extreme precipitation along the equator did not exhibit a strong correlation with ARs due to the characteristic elongation of ARs from equatorial regions toward mid-latitudes.

The geographical distribution of longer AR durations and concurrent extreme precipitation clustered notably along North America, South America, Europe, Central East Asia, and Australia. Moreover, grids with AR durations exceeding 12 h emerged as key contributors to precipitation extremes, showcasing importance of prolonged AR activity in generating extreme weather conditions (Fig. 4.2).

Further analysis revealed that areas experiencing longer AR durations and elevated extreme precipitation levels often exhibited a similar pattern with IVT magnitudes. This parallel distribution pattern indicated a close relationship between AR intensity, duration, and the resulting extreme precipitation, underscoring the complex interplay between atmospheric moisture transport by ARs and the occurrence of extreme weather events.

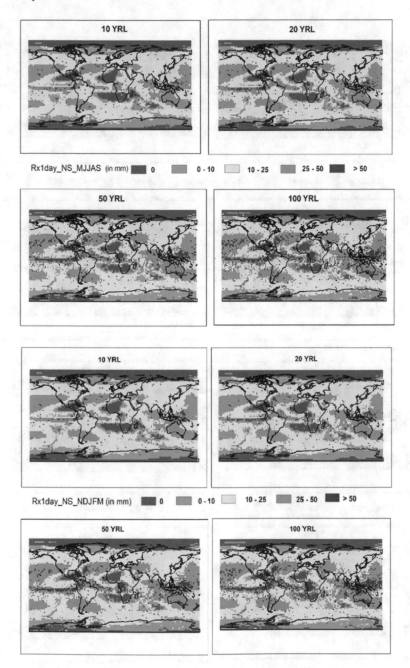

(a) Rx1day for seasons MJJAS and NDJFM

Fig. 4.1 Return levels of precipitation extremes **a** Rx1day, **b** Rx5day, and **c** R95pTOT for MJJAS and NDJFM climatology at 10, 20, 50, and 100 years

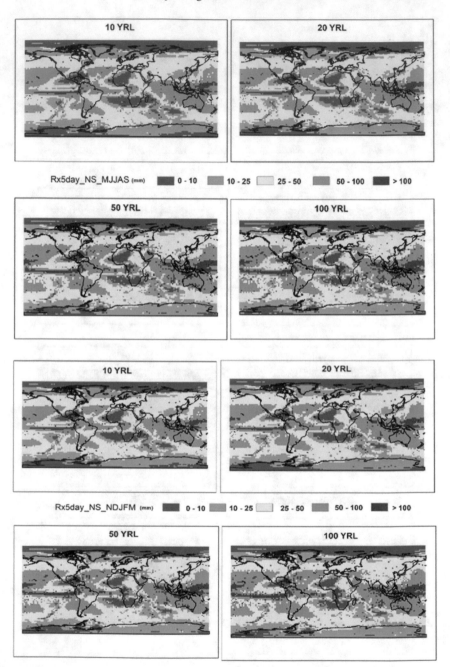

(b) Rx5day for seasons MJJAS and NDJFM

Fig. 4.1   (continued)

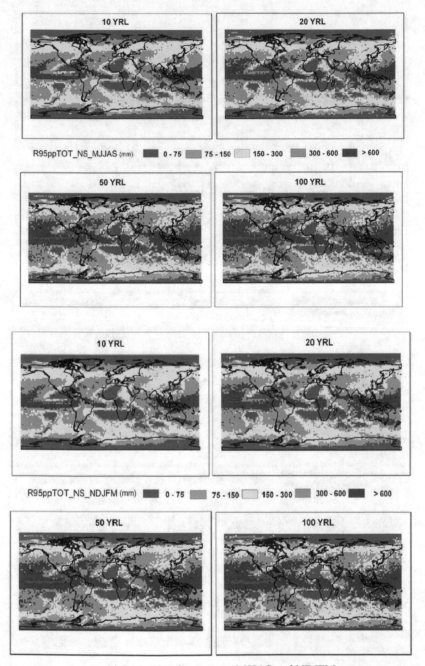

(c) R95pTOT for seasons MJJAS and NDJFM

Fig. 4.1 (continued)

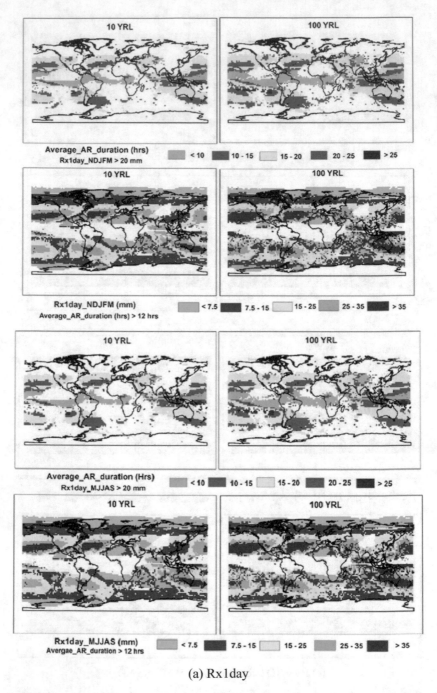

(a) Rx1day

Fig. 4.2 Spatial distribution of annual average AR duration under different return levels and conditions for both seasons. The upper half shows results for Rx1day > 20 mm, Rx5day > 30 mm, and R95pTOT > 200 mm over 10-year and 100-year return periods. The lower half presents the same return levels but with average AR durations exceeding 12 h

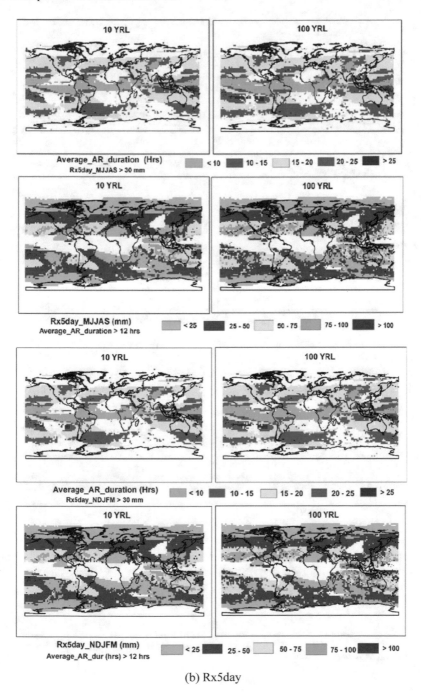

(b) Rx5day

Fig. 4.2 (continued)

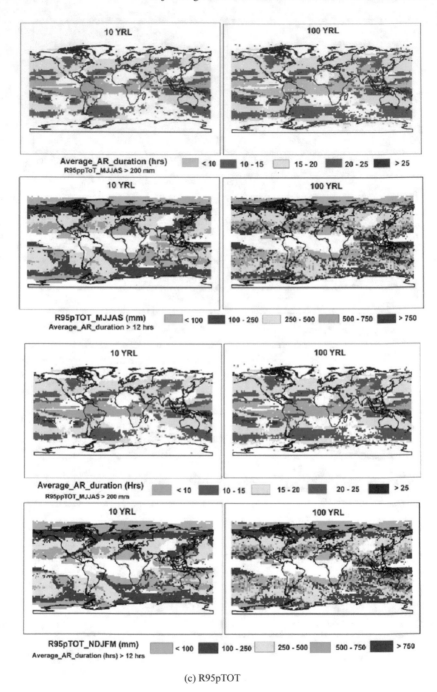

(c) R95pTOT

Fig. 4.2 (continued)

4.3.2 Precipitation Extremes and AR Frequency

The analysis also examined the relationship between precipitation extremes and AR frequency using specific thresholds (Rx1day > 20 mm, Rx5day > 30 mm, and R95pTOT > 200 mm). The proportion of timesteps during which a grid cell is characterized as an AR, relative to the total number of timesteps in percentage, is considered as AR frequency.

Findings from the study revealed that spatial extent of high AR frequencies corresponded to high magnitude of precipitation extremes. Specifically, in Northern Hemisphere, higher AR frequency and associated return levels were observed along North America and Europe during the NDJFM climatology. Conversely, in Southern Hemisphere, more AR frequency and high return levels were seen over southernmost South America and Australia. This pattern indicated a strong correlation between AR frequency and precipitation extremes. Furthermore, the analysis showed that higher return levels of extreme precipitation, when combined with seasonal covariates (NDJFM and MJJAS), exhibited a similar spatial distribution to AR frequency. Additionally, both the magnitude and spatial extent of precipitation extremes correlated with high AR frequency increased with longer periods (10 or 100 years) under consideration, highlighting the evolving nature of extreme precipitation patterns linked to AR activity over time (Fig. 4.3).

4.4 Future Research and Climate Adaptation

ARs, influenced by climate oscillations, are crucial components of the global water cycle, replenishing water resources and supporting ecosystems. However, extreme AR events can trigger water-related disasters posing challenges to human settlements and infrastructure resilience. To balance the beneficial aspects of ARs with their potential hazards, integrated water management approaches, sustainable resource utilization, and adaptive ecosystem strategies are essential.

Continued research on climate oscillations and their interactions with ARs is imperative for enhancing predictive capabilities, refining climate modeling accuracy, and developing robust climate adaptation strategies. Collaborative efforts among scientists, policymakers, and stakeholders are crucial for implementing proactive measures to address climate variability, mitigate risks associated with extreme weather events, and promote climate resilience in vulnerable regions. Embracing a holistic approach to climate research and adaptation is key to fostering sustainable development and environmental stewardship in a changing climate landscape.

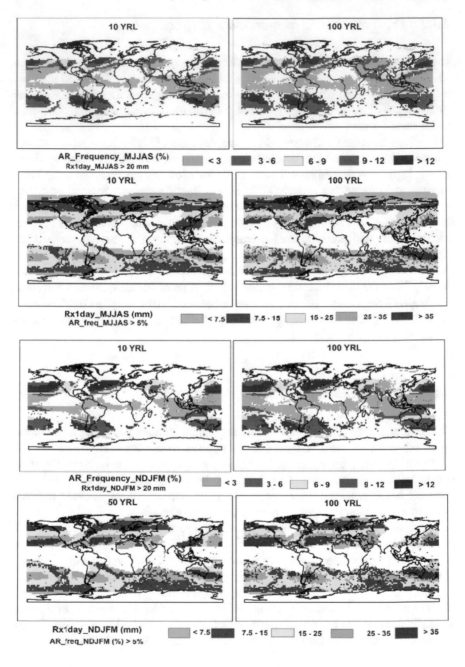

(a) Rx1day

Fig. 4.3 Spatial distribution of average seasonal (MJJAS and NDJFM) AR frequency under various return levels and conditions. The upper half depicts results for Rx1day > 20 mm, Rx5day > 30 mm, and R95pTOT > 200 mm over 10-year and 100-year return periods. The lower half shows these return levels with seasonal AR frequencies surpassing 5% for both seasons

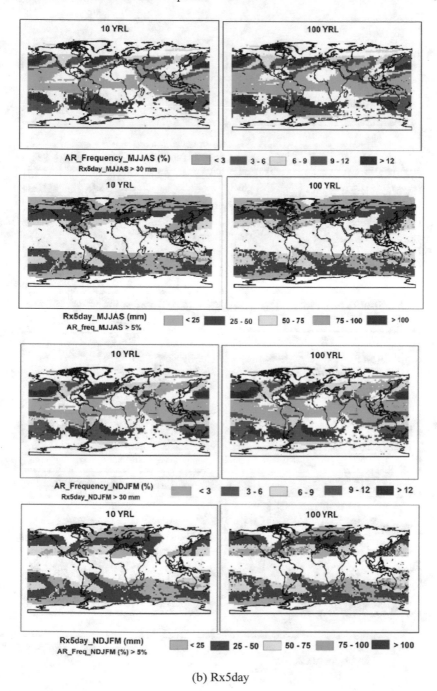

(b) Rx5day

Fig. 4.3 (continued)

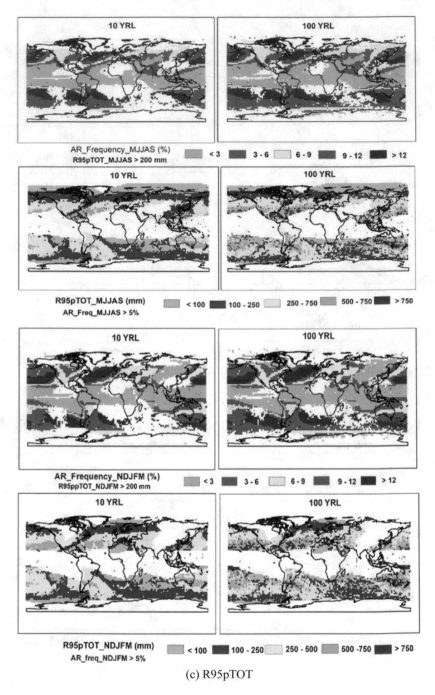

(c) R95pTOT

Fig. 4.3 (continued)

4.5 Conclusions

Climate oscillations exert a profound influence on global weather systems and regional hydrology, with their intricate dynamics interacting closely with atmospheric rivers (ARs). These interactions have substantial implications for climate variability, extreme weather occurrences, and the management of water resources. Notably, Large-Scale Climate Oscillations (LSCOs) like ENSO, NAO, and AO are pivotal in shaping these phenomena, showing clear connections to AR behavior in regions such as North America and Europe. For example, shifts in AR frequency during El Niño phases illustrate how LSCOs can modulate AR dynamics, affecting different regions in diverse ways and highlighting the complexity of their relationships.

This study conducted a detailed analysis that compared stationary and nonstationary models using statistical techniques, focusing on key factors like AO, NAO, ENSO, and PDO. The findings underscored the significant impact of PDO, a long-lasting ENSO phenomenon, on large-scale weather patterns, especially evident during the summer months. Additionally, the study revealed that seasonal PDO indices, particularly those for MJJAS, were more effective in capturing variations in extreme precipitation events compared to other indices, indicating a strong influence of PDO on precipitation patterns in specific regions.

A better knowledge of these interconnections is important for improving the accuracy of AR predictions across different time scales and for developing effective strategies to manage risks related to climate extremes. Collaborative efforts and advancements in scientific research play essential roles in this endeavor, as they contribute to better forecasting capabilities, water resource management practices, and enhanced resilience with evolving climate conditions. The insights of study into the complex interplay between climate oscillations, ARs, and precipitation extremes provide a valuable foundation for refining risk assessment methodologies and implementing targeted management strategies in vulnerable regions worldwide.

References

1. V. Agilan, N.V. Umamahesh, What are the best covariates for developing non-stationary rainfall intensity-duration-frequency relationship? Adv. Water Resour. **101**, 11–22 (2017). https://doi.org/10.1016/j.advwatres.2016.12.016
2. V. Agilan, N.V. Umamahesh, El Niño Southern Oscillation cycle indicator for modeling extreme rainfall intensity over India. Ecol. Ind. **84**, 450–458. https://doi.org/10.1016/j.ecolind.2017.09.012
3. D. Baldan, E. Coraci, F. Crosato, M. Ferla, A. Bonometto, S. Morucci, Importance of non-stationary analysis for assessing extreme sea levels under sea level rise. Nat. Hazard. **22**(11), 3663–3677 (2022). https://doi.org/10.5194/nhess-22-3663-2022
4. C. Bracken, K.D. Holman, B. Rajagopalan, H. Moradkhani, A Bayesian hierarchical approach to multivariate nonstationary hydrologic frequency analysis. Water Resourc. Res. **54**(1), 243–255 (2018). https://doi.org/10.1002/2017WR020403
5. S. Coles, *An Introduction to Statistical Modeling of Extreme Values* (Springer, 2001)

6. J. Das, S. Jha, M.K. Goyal, On the relationship of climatic and monsoon teleconnections with monthly precipitation over meteorologically homogenous regions in India: wavelet & global coherence approaches. Atmos. Res. **238**, 104889 (2020). https://doi.org/10.1016/j.atmosres.2020.104889

7. J. Das, S. Jha, M.K. Goyal, Non-stationary and copula-based approach to assess the drought characteristics encompassing climate indices over the Himalayan states in India. J. Hydrol. **580**, 124356 (2020). https://doi.org/10.1016/j.jhydrol.2019.124356

8. C. Deser, On the teleconnectivity of the "Arctic oscillation". 779–782 (2000)

9. A. Gershunov, T. Shulgina, R.E.S. Clemesha, K. Guirguis, D.W. Pierce, M.D. Dettinger, D.A. Lavers, D.R. Cayan, S.D. Polade, J. Kalansky, F.M. Ralph, Precipitation regime change in Western North America: the role of atmospheric rivers. Sci. Rep. **9**(1), 9944 (2019). https://doi.org/10.1038/s41598-019-46169-w

10. B. Guan, N.P. Molotch, D.E. Waliser, E.J. Fetzer, P.J. Neiman, The 2010/2011 snow season in California's Sierra Nevada: role of atmospheric rivers and modes of large-scale variability. Water Resour. Res. **49**(10), 6731–6743 (2013). https://doi.org/10.1002/wrcr.20537

11. B. Guan, D.E. Waliser, Detection of atmospheric rivers: Evaluation and application of an algorithm for global studies. J. Geophys. Res. Atmos. **120**(24), 12514–12535 (2015). https://doi.org/10.1002/2015JD024257

12. K. Guirguis, A. Gershunov, R.E.S. Clemesha, T. Shulgina, A.C. Subramanian, F.M. Ralph, Circulation drivers of atmospheric rivers at the North American West Coast. Geophys. Res. Lett. **45**, 12576–12584 (2018). https://doi.org/10.1029/2018GL079249

13. X. Guo, N. Zhao, K. Kikuchi, T. Nasuno, M. Nakano, H. Annamalai, Atmospheric rivers over the indo-pacific and its associations with the boreal summer intraseasonal oscillation. J. Clim. **34**(24), 9711–9728 (2021). https://doi.org/10.1175/JCLI-D-21-0290.1

14. S. Jha, J. Das, M.K. Goyal, Low frequency global-scale modes and its influence on rainfall extremes over India: nonstationary and uncertainty analysis. Int. J. Climatol. **41**(3), 1873–1888 (2021). https://doi.org/10.1002/joc.6935

15. S. Jha, M.K. Goyal, B.B. Gupta, C. Hsu, E. Gilleland, J. Das, A methodological framework for extreme climate risk assessment integrating satellite and location based data sets in intelligent systems. Int. J. Intell. Syst. **37**(12), 10268–10288 (2022). https://doi.org/10.1002/int.22475

16. R.W. Katz, Statistical methods for nonstationary extremes. Extremes Changing Clim. 15–37 (2013)

17. H. Kim, S. Kim, H. Shin, J.-H. Heo, Appropriate model selection methods for nonstationary generalized extreme value models. J. Hydrol. **547**, 557–574 (2017). https://doi.org/10.1016/j.jhydrol.2017.02.005

18. N. Kumar, P. Patel, S. Singh, M.K. Goyal, Understanding non-stationarity of hydroclimatic extremes and resilience in Peninsular catchments, India. Sci. Rep. **13**(1), 12524 (2023). https://doi.org/10.1038/s41598-023-38771-w

19. D.A. Lavers, G. Villarini, The nexus between atmospheric rivers and extreme precipitation across Europe. Geophys. Res. Lett. **40**(12), 3259–3264 (2013). https://doi.org/10.1002/grl.50636

20. M. Ma, M. Ren, H. Zang, H. Cui, S. Jiang, Y. Sun, Nonstationary quantity-duration-frequency (QDF) relationships of lowflow in the source area of the Yellow River basin, China. J. Hydrol. Reg. Stud. **48**, 101450 (2023). https://doi.org/10.1016/j.ejrh.2023.101450

21. S. Meghani, S. Singh, N. Kumar, M.K. Goyal, Predicting the spatiotemporal characteristics of atmospheric rivers: a novel data-driven approach. Global Planet. Change **231**, 104295 (2023). https://doi.org/10.1016/j.gloplacha.2023.104295

22. A. Mondal, P.P. Mujumdar, Modeling non-stationarity in intensity, duration and frequency of extreme rainfall over India. J. Hydrol. **521**, 217–231 (2015). https://doi.org/10.1016/j.jhydrol.2014.11.071

23. P.J. Neiman, L.J. Schick, F.M. Ralph, M. Hughes, G.A. Wick, Flooding in Western Washington: the connection to atmospheric rivers. J. Hydrometeorol. **12**(6), 1337–1358 (2011). https://doi.org/10.1175/2011JHM1358.1

24. T.B.M.J. Ouarda, C. Charron, Changes in the distribution of hydro-climatic extremes in a nonstationary framework. Sci. Rep. **9**(1), 8104 (2019). https://doi.org/10.1038/s41598-019-446 03-7

25. H. Paltan, D. Waliser, W.H. Lim, B. Guan, D. Yamazaki, R. Pant, S. Dadson, Global floods and water availability driven by atmospheric rivers. Geophys. Res. Lett. **44**(20), 10387–10395 (2017). https://doi.org/10.1002/2017GL074882

26. S.D. Polade, A. Gershunov, D.R. Cayan, M.D. Dettinger, D.W. Pierce, Natural climate variability and teleconnections to precipitation over the Pacific-North American region in CMIP3 and CMIP5 models. Geophys. Res. Lett. **40**(10), 2296–2301 (2013). https://doi.org/10.1002/grl.50491

27. K.S. Rautela, S. Singh, M.K. Goyal, Characterizing the spatio-temporal distribution, detection, and prediction of aerosol atmospheric rivers on a global scale. J. Environ. Manage. **351**, 119675 (2024). https://doi.org/10.1016/j.jenvman.2023.119675

28. P. Shi, T. Yang, C.-Y. Xu, Y. Yong, Q. Shao, Z. Li, X. Wang, X. Zhou, S. Li, How do the multiple large-scale climate oscillations trigger extreme precipitation? Global Planet. Change **157**, 48–58 (2017). https://doi.org/10.1016/j.gloplacha.2017.08.014

29. S. Singh, M.K. Goyal, Enhancing climate resilience in businesses: the role of artificial intelligence. J. Cleaner Prod. 138228 (2023)

30. S. Singh, M.K. Goyal, An innovative approach to predict atmospheric rivers: Exploring convolutional autoencoder. Atmos. Res. **289**, 106754 (2023)

31. S. Singh, M.K. Goyal, S. Jha, Role of large-scale climate oscillations in precipitation extremes associated with atmospheric rivers: nonstationary framework. Hydrol. Sci. J. **68**(3), 395–411 (2023)

32. S. Singh, N. Kumar, M.K. Goyal, S. Jha, Relative influence of ENSO, IOD, and AMO over spatiotemporal variability of hydroclimatic extremes in Narmada basin, India. AQUA—Water Infrastruct. Ecosyst. Soc.**72**(4), 520–539. https://doi.org/10.2166/aqua.2023.219

33. S. Singh, A. Yadav, G.M. Kumar, Univariate and bivariate spatiotemporal characteristics of heat waves and relative influence of large-scale climate oscillations over India. J. Hydrol. **628**, 130596 (2024). https://doi.org/10.1016/j.jhydrol.2023.130596

34. Z. Song, J. Xia, D. She, L. Zhang, C. Hu, L. Zhao, The development of a nonstationary standardized precipitation index using climate covariates: a case study in the middle and lower reaches of Yangtze River Basin, China. J. Hydrol. **588**, 125115 (2020). https://doi.org/10.1016/j.jhydrol.2020.125115

35. S. Sugahara, R.P. da Rocha, R. Silveira, Non-stationary frequency analysis of extreme daily rainfall in Sao Paulo, Brazil. Int. J. Climatol. **29**(9), 1339–1349 (2009). https://doi.org/10.1002/joc.1760

36. P.J. Ward, S. Eisner, M. Flörke, M.D. Dettinger, M. Kummu, Annual flood sensitivities to El Niño-Southern Oscillation at the global scale. Hydrol. Earth Syst. Sci. **18**(1), 47–66 (2014). https://doi.org/10.5194/hess-18-47-2014

37. S. Wi, J. Valdes, S. Steinschneider, T.-W. Kim, Non-stationary frequency analysis of extreme precipitation in South Korea using peaks-over-threshold and annual maxima. Stochast. Environ. Res. Risk Assess. **30** (2016). https://doi.org/10.1007/s00477-015-1180-8

38. P. Xu, Y. Wang, X. Fu, V.P. Singh, J. Qiu, Detection and attribution of urbanization impact on summer extreme heat based on nonstationary models in the Yangtze River Delta, China. Urban Clim. **47**, 101376 (2023). https://doi.org/10.1016/j.uclim.2022.101376

39. R. Yao, S. Zhang, P. Sun, Q. Dai, Q. Yang, Estimating the impact of urbanization on nonstationary models of extreme precipitation events in the Yangtze River Delta metropolitan region. Weather Clim. Extremes **36**, 100445 (2022). https://doi.org/10.1016/j.wace.2022.100445

40. A.G. Yilmaz, B.J.C. Perera, Extreme rainfall nonstationarity investigation and intensity–frequency–duration relationship. J. Hydrol. Eng. **19**(6), 1160–1172 (2014). https://doi.org/10.1061/(ASCE)HE.1943-5584.0000878

Chapter 5
Role of Machine Learning in Understanding and Managing Atmospheric Rivers

5.1 Introduction

Recognizing the pivotal role of Atmospheric Rivers (ARs) in both global and regional climatology, significant attention has been directed toward monitoring and forecasting their trajectory and key characteristics upon landfall [10, 11, 31, 34]. Accurate predictions of AR attributes with extended lead times are highly valuable, especially for effective flood disaster management and water resource planning. Traditionally, weather and climate forecasting rely heavily on Numerical Weather Prediction (NWP) models [20, 28]. These models integrate fundamental physical processes governing the atmosphere, oceans, and land, incorporating principles derived from mass, energy, and momentum conservation laws [4, 7, 18, 22, 30, 32, 35]. However, inherent uncertainties arising from numerical approximations, model initialization conditions, and inherent model constraints gradually erode the prediction accuracy over time [3, 8, 15, 16, 29, 37, 38]. Studies evaluating ensemble forecast systems in the Northeast Pacific Ocean region revealed the capability of the models to detect the presence of ARs. However, these models faced challenges in accurately predicting the finer details of landfalling ARs at significant lead times [23, 24, 39]. Similarly, Nardi et al. [20] highlighted substantial forecasting errors in NWP models specifically related to AR landfall predictions. As a result, endeavors have been concentrated on improving data collection, highlighting crucial atmospheric structures, and seamlessly integrating this information into NWP models to enhance forecasting accuracy [6, 40].

Artificial Intelligence, a rapidly evolving field of computer science, empowers machines to learn, analyze data, and make decisions without explicit programming [3, 9, 12]. Its applications span various domains, leveraging algorithms, machine learning, and neural networks to process large datasets, identify patterns, and derive insights that traditional methods might overlook [1, 14, 36]. The adaptability and capacity of AI to handle complex data make it a potent tool in addressing multifaceted challenges [21, 25, 37]. In recent years, AI has emerged as a game-changer in

© The Author(s), under exclusive license to Springer Nature Switzerland AG 2024   67
M. K. Goyal and S. Singh, *Understanding Atmospheric Rivers Using Machine Learning*,
SpringerBriefs in Applied Sciences and Technology,
https://doi.org/10.1007/978-3-031-63478-9_5

atmospheric sciences, particularly in understanding and managing climate extremes [17]. AI-powered models and algorithms can contribute significantly to AR prediction accuracy, enabling more precise forecasting of AR occurrences, trajectories, and intensities. These advanced prediction capabilities are essential in issuing timely warnings, thereby mitigating the potential impacts of AR-induced extreme weather events. Taking inspiration from the capacity of Deep Learning (DL) architectures to comprehend intricate and nonlinear relationships, exploring DL algorithms offers an alternative way to grasp the dynamics inherent in ARs. Unlike NWP models, these data-driven neural networks can discern atmospheric conditions over time by analyzing the given atmospheric data, bypassing the necessity for explicitly modeling physical processes [7].

5.2 Potential Applications of AI in Understanding ARs

Artificial Intelligence (AI) plays a pivotal role in understanding and dealing with ARs, offering innovative solutions in several ways:

(a) **Pattern Recognition**: Recent research has showcased that Machine Learning (ML) algorithms provide better results in recognizing intricate patterns within vast and complex datasets. When applied to atmospheric data, AI can enable the identification of intricate features and correlations associated with ARs. It can discern subtle variations in atmospheric variables like moisture content, wind behavior, temperature gradients, and pressure systems that characterize ARs. By analyzing historical and real-time data, AI can help identify the specific signatures indicative of AR formation, evolution, and behavior. This detailed pattern recognition enhances our comprehension of how ARs develop, intensify, and interact with other weather phenomena, contributing to more accurate tracking and forecasting.

(b) **Prediction and Modeling**: AI-powered models, such as deep learning algorithms and neural networks, revolutionize the prediction and modeling of ARs. These models handle large volumes of atmospheric data, learning intricate relationships and nonlinear patterns that traditional models might miss. Through continuous learning from data, AI models enhance the precision of AR forecasts, predicting their paths, intensities, and durations with higher accuracy. Moreover, these AI-based predictive models adapt and improve over time, incorporating new data and refining their predictions, thus bolstering our ability to anticipate and prepare for AR-related weather events.

(c) **Early Warning Systems**: AI-driven early warning systems can be pivotal in mitigating the impacts of AR-related extreme weather events. By integrating AI algorithms with diverse data sources like satellite imagery, radar data, and atmospheric sensors, these systems can rapidly identify and track the onset and movement of ARs. AI algorithms can analyze the evolving atmospheric conditions, issuing timely alerts and forecasts to relevant authorities and communities.

These warnings can provide essential lead times, enabling proactive measures to mitigate risks associated with ARs, such as flooding, heavy rainfall, or landslides. As a result, AI-powered early warning systems contribute significantly to minimizing damage and enhancing disaster preparedness.

5.3 Case Study: AR Prediction Using Deep Learning Model

Motivated by the adeptness of DL models in unraveling intricate patterns and delivering precise predictions, this research delved into employing DL techniques to comprehend the dynamics of Atmospheric Rivers (ARs). DL methods outperform typical NWP models in capturing spatiotemporal characteristics in atmospheric data without explicitly including physical processes [5, 27]. The study utilized ERA5 reanalysis data (specific humidity, zonal and meridional wind velocities) from the European Centre for Medium-Range Weather Forecasts (ECMWF) to characterize AR states. Following the methodology of [11], IVT characteristics were employed to delineate ARs at each timestep (Fig. 5.1). The study employed an autoencoder DL model to forecast the presence of ARs in upcoming timesteps. The insights from the research article offer a new perspective for atmospheric researchers, highlighting the skill of DL models in enhancing the precision and reliability of AR prediction.

5.3.1 AR Prediction

Leveraging the capabilities of convolutional autoencoders to capture both spatial and temporal patterns, our approach involved using these models to characterize the attributes of ARs. To train our DL model, we employed AR maps detected at 6-h intervals as input. These frames, depicting the presence of ARs, served as the foundation for our modeling process. The convolutional layers in the proposed Autoencoder architecture play an important role in extracting spatial information

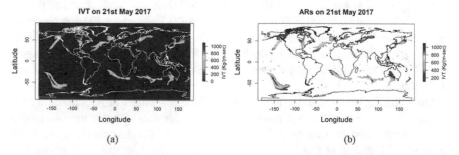

Fig. 5.1 (A) IVT characteristics on an arbitrary timestep in May 21, 2017 and (b) Detected ARs after characterizing the intense atmospheric field of integrated water vapor transport

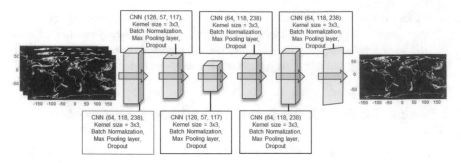

Fig. 5.2 The configuration of the proposed convolutional autoencoder

from the ARs frames. and this information was then encoded into a representation within a latent space. The standard configuration of the model comprised: an encoder, a bottleneck, and a decoder [13, 26] (Fig. 5.2).

The input data was prepossessed in a sequence of 11 consecutive images to train the proposed DL model for predicting the 11th AR frame using sequences of 10 AR frames. The objective was to minimize the training loss and further enhance model performance by optimizing the architecture. Additionally, an L2 loss was used as a reconstruction loss. The input data was processed in batches of 10 frames during training. Despite observing an increase in training loss after 23 epochs, we continued training for 30 epochs to ensure comprehensive learning. During the training period, the model achieved a minimum mean squared error loss of 0.025, taking approximately 61 min to complete 30 epochs.

To verify the predictability of the model, commonly used metrics such as Structural Similarity Index (SSIM) and Peak Signal-to-Noise Ratio (PSNR) were used. PSNR assesses picture quality by comparing original and compressed images, while SSIM evaluates the similarity between predicted and original images. Reduced MSE and RMSE alongside elevated PSNR and SSIM indicate the improved predictive capabilities of the proposed model [33].

An illustrative example in Fig. 5.3 showcases the predictive capabilities of the model, depicting the reconstruction of the next from the testing dataset, compared to the original. The evaluation yielded on average RMSE and MSE scores of 0.1552 and 0.0247, along with SSIM and PSNR scores of 0.739 and 64.421, respectively, on the testing dataset [19]. These improved metric scores signify superior prediction performance compared to a related study by [2], focusing on predicting urban expansion using a Conv-LSTM model used similar metrics to evaluate the predictability of the model.

Modeling the intricate and dynamic spatiotemporal features of ARs poses significant challenges, especially in accurately predicting them using NWP models, particularly for extended lead times. This often leads to considerable inaccuracies, especially when forecasting the landfall locations [31]. The potential of sophisticated

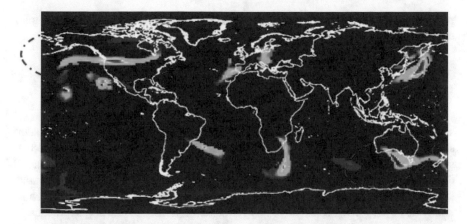

(a) AR Observed

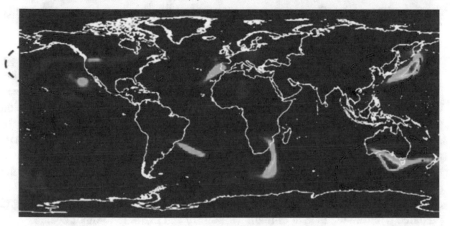

(b) AR Predicted

Fig. 5.3 Illustrating the prediction of AR availability through proposed autoencoder: **a** ARs observed at an arbitrary timestep and **b** ARs predicted by the trained deep learning model at same time instance

deep learning models lies in their ability to capture nonlinear and intricate inter-actions within datasets, providing a viable solution for understanding the dynamic behavior of ARs.

Encouraged by advancements in DL models and transfer learning methods, we were motivated to investigate DL algorithms for capturing spatiotemporal features of ARs. Through optimization of training of our model guided by insights from relevant studies, we attained promising outcomes in AR predictability. The integration of AI in dealing with ARs showcases the immense potential for advancing our understanding and prediction capabilities, and mitigating the impacts of these impactful weather phenomena. It stands as a testament to the power of technology in addressing complex climate-related challenges.

5.4 Conclusions

The integration of AI has emerged as a transformative force in an approach to understanding and managing ARs. The capacity of AI to process vast amounts of atmospheric data has proven instrumental in uncovering intricate patterns inherent in AR dynamics. This analytical prowess facilitates the identification of crucial AR characteristics, enabling refined classification, tracking, and comprehension of their behaviors. One of the pivotal roles played by AI lies in its ability to develop sophisticated risk assessment tools tailored to evaluate vulnerabilities within regions prone to AR-induced hazards. By discerning these vulnerabilities, AI-driven analyses empower the formulation of resilient strategies and adaptive measures. This proactive stance becomes crucial in mitigating potential impacts and bolstering the resilience of communities and infrastructure against AR-related adversities.

The integration of AI into AR research and management holds immense promise. It not only advances our comprehension of these inherently complex phenomena but also stands as a proactive force in mitigating their adverse effects. Looking ahead, the evolution of AI technologies holds vast potential. Innovations in data fusion techniques, the establishment of robust early warning systems, and the application of AI-driven climate change analyses promise to revolutionize our ability to address AR-related challenges. These advancements are poised to redefine disaster preparedness strategies and fortify climate resilience in vulnerable regions. By harnessing the power of AI, we move closer to a future where communities are better equipped to anticipate, mitigate, and adapt to the impacts of ARs, fostering a more resilient and prepared world in the face of climatic variability and extreme weather events.

References

1. R. Ben Ayed, M. Hanana, Artificial intelligence to improve the food and agriculture sector. J. Food Qual. **2021**, 1–7 (2021). https://doi.org/10.1155/2021/5584754
2. W. Boulila, H. Ghandorh, M.A. Khan, F. Ahmed, J. Ahmad, A novel CNN-LSTM-based approach to predict urban expansion. Eco. Inform. **64**, 101325 (2021). https://doi.org/10.1016/j.ecoinf.2021.101325
3. Chandrrasekar, B.R. Gurjar, C.S.P. Ojha, M.K. Goyal, Closure to "Potential assessment of neural network and decision tree algorithms for forecasting ambient PM2.5and CO concentrations: case study" by Chandrra Sekar, B. R. Gurjar, C. S. P. Ojha, and Manish Kumar Goyal. J. Hazardous, Toxic, Radioactive Waste **21**(4), 1–9 (2017). https://doi.org/10.1061/(ASCE)HZ. 2153-5515.0000276
4. W.E. Chapman, A.C. Subramanian, L. Delle Monache, S.P. Xie, F.M. Ralph, Improving atmospheric river forecasts with machine learning. Geophys. Res. Lett. **46**(17–18), 10627–10635 (2019). https://doi.org/10.1029/2019GL083662

5. S. Chattopadhyay, G. Bandyopadhyay, Artificial neural network with backpropagation learning to predict mean monthly total ozone in Arosa, Switzerland. Int. J. Remote Sens. **28**(20), 4471–4482 (2007). https://doi.org/10.1080/01431160701250440

6. A. Cobb, F.M. Ralph, V. Tallapragada, A.M. Wilson, C.A. Davis, L.D. Monache, J.D. Doyle, F. Pappenberger, C.A. Reynolds, A. Subramanian, P.G. Black, F. Cannon, C. Castellano, J.M. Cordeira, J.S. Haase, C. Hecht, B. Kawzenuk, D.A. Lavers, M.J. Murphy, J. Parrish, R. Rickert, J.J. Rutz, R. Torn, X. Wu, M. Zheng, Atmospheric river reconnaissance 2021: a review. Weather Forecast. (2022). https://doi.org/10.1175/WAF-D-21-0164.1

7. P.D. Dueben, P. Bauer, Challenges and design choices for global weather and climate models based on machine learning. Geosci Model Develop. **11**(10), 3999–4009 (2018). https://doi.org/10.5194/gmd-11-3999-2018

8. M.K. Goyal, B. Bharti, J. Quilty, J. Adamowski, A. Pandey, Modeling of daily pan evaporation in sub tropical climates using ANN, LS-SVR, Fuzzy Logic, and ANFIS. Expert Syst. Appl. **41**(11), 5267–5276 (2014). https://doi.org/10.1016/j.eswa.2014.02.047

9. M.K. Goyal, B. Bharti, J. Quilty, J. Adamowski, A. Pandey, Modeling of daily pan evaporation in sub tropical climates using ANN, LS-SVR, Fuzzy Logic, and ANFIS. Expert Syst. Appl. **41**(11), 5267–5276 (2014). https://doi.org/10.1016/j.eswa.2014.02.047

10. B. Guan, D.E. Waliser, Atmospheric rivers in 20 year weather and climate simulations: a multimodel, global evaluation. J. Geophys. Res. Atmos. **122**(11), 5556–5581 (2017). https://doi.org/10.1002/2016JD026174

11. B. Guan, D.E. Waliser, Tracking atmospheric rivers globally: Spatial distributions and temporal evolution of life cycle characteristics. J. Geophys. Res. Atmos. **124**, 12523–12552 (2019). https://doi.org/10.1029/2019JD031205

12. G. Hinge, R.Y. Surampalli, M.K. Goyal, Prediction of soil organic carbon stock using digital mapping approach in humid India. Environ. Earth Sci. **77**(5), 172 (2018). https://doi.org/10.1007/s12665-018-7374-x

13. D. Jana, J. Patil, S. Herkal, S. Nagarajaiah, L. Duenas-Osorio, CNN and convolutional autoencoder (CAE) based real-time sensor fault detection, localization, and correction. Mech. Syst. Signal Process. **169**, 108723 (2022). https://doi.org/10.1016/j.ymssp.2021.108723

14. B. Jena, S. Saxena, G.K. Nayak, L. Saba, N. Sharma, J.S. Suri, Artificial intelligence-based hybrid deep learning models for image classification: the first narrative review. Comput. Biol. Med. **137**, 104803 (2021). https://doi.org/10.1016/j.compbiomed.2021.104803

15. S. Jha, M.K. Goyal, B.B. Gupta, C. Hsu, E. Gilleland, J. Das, A methodological framework for extreme climate risk assessment integrating satellite and location based data sets in intelligent systems. Int. J. Intell. Syst. **37**(12), 10268–10288 (2022). https://doi.org/10.1002/int.22475

16. M. Krishan, S. Jha, J. Das, A. Singh, M.K. Goyal, C. Sekar, Air quality modelling using long short-term memory (LSTM) over NCT-Delhi, India. Air Qual. Atmos. Health. 899–908 (2019). https://doi.org/10.1007/s11869-019-00696-7

17. D.N. Kumar, M.J. Reddy, R. Maity, Regional rainfall forecasting using large scale climate teleconnections and artificial intelligence techniques. J. Intell. Syst. **16**(4) (2007). https://doi.org/10.1515/JISYS.2007.16.4.307

18. N. Kumar, P. Patel, S. Singh, M.K. Goyal, Understanding non-stationarity of hydroclimatic extremes and resilience in Peninsular catchments, India. Sci. Rep. **13**(1), 12524 (2023). https://doi.org/10.1038/s41598-023-38771-w

19. S. Meghani, S. Singh, N. Kumar, M.K. Goyal, Predicting the spatiotemporal characteristics of atmospheric rivers: a novel data-driven approach. Global Planet. Change **231**, 104295 (2023). https://doi.org/10.1016/j.gloplacha.2023.104295

20. K.M. Nardi, E.A. Barnes, F.M. Ralph, Assessment of numerical weather prediction model reforecasts of the occurrence, intensity, and location of atmospheric rivers along the west coast of North America. Mon. Weather Rev. **146**(10), 3343–3362 (2018). https://doi.org/10.1175/MWR-D-18-0060.1

21. R. Nishant, M. Kennedy, J. Corbett, Artificial intelligence for sustainability: challenges, opportunities, and a research agenda. Int. J. Inf. Manage. **53**, 102104 (2020). https://doi.org/10.1016/j.ijinfomgt.2020.102104

22. S. Rakkasagi, M.K. Goyal, S. Jha, Evaluating the future risk of coastal Ramsar wetlands in India to extreme rainfalls using fuzzy logic. J. Hydrol. **632**, 130869 (2024). https://doi.org/10.1016/j.jhydrol.2024.130869

23. K.S. Rautela, D. Kumar, B.G.R. Gandhi, A. Kumar, A.K. Dubey, Application of ANNs for the modeling of streamflow, sediment transport, and erosion rate of a high-altitude river system in Western Himalaya, Uttarakhand. RBRH**27** (2022). https://doi.org/10.1590/2318-0331.272220220045

24. K.S. Rautela, J.C. Kuniyal, M.K. Goyal, N. Kanwar, A.S. Bhoj, Assessment and modelling of hydro-sedimentological flows of the eastern river Dhauliganga, north-western Himalaya, India. Nat.ural Hazards (2024). https://doi.org/10.1007/s11069-024-06413-7

25. K.S. Rautela, S. Singh, M.K. Goyal, Characterizing the spatio-temporal distribution, detection, and prediction of aerosol atmospheric rivers on a global scale. J. Environ. Manage. **351**, 119675 (2024). https://doi.org/10.1016/j.jenvman.2023.119675

26. D.E. Rumelhart, G.E. Hinton, R.J. Williams, Learning representations by back-propagating errors. Nature **323**(6088), 533–536 (1986). https://doi.org/10.1038/323533a0

27. S. Scher. Toward data-driven weather and climate forecasting: approximating a simple general circulation model with deep learning. Geophys. Res. Lett. **45**(22), 12,612–616,622 (2018). https://doi.org/10.1029/2018GL080704

28. S. Scher, G. Messori, Predicting weather forecast uncertainty with machine learning. Quart. J. Roy. Meteorol. Soc. **144**(717), 2830–2841 (2018). https://doi.org/10.1002/qj.3410

29. A. Sharma, M.K. Goyal, A comparison of three soft computing techniques, Bayesian regression, support vector regression, and wavelet regression, for monthly rainfall forecast. J. Intell. Syst. **26**(4), 641–655 (2017). https://doi.org/10.1515/jisys-2016-0065

30. A. Sharma, M.K. Goyal, Assessment of drought trend and variability in India using wavelet transform. Hydrol. Sci. J. **65**(9), 1539–1554 (2020). https://doi.org/10.1080/02626667.2020.1754422

31. C.A. Shields, J.J. Rutz, L.R. Leung, F.M. Ralph, M. Wehner, T. O'Brien, Defining uncertainties through comparison of atmospheric river tracking methods. Bull. Am. Meteorol. Soc. **100**, ES93–ES96 (2019). https://doi.org/10.1175/BAMS-D-18-0200.1

32. S. Singh, M.K. Goyal, An innovative approach to predict atmospheric rivers: exploring convolutional autoencoder. Atmos. Res. **289**, 106754 (2023)

33. S. Singh, M.K. Goyal, Enhancing climate resilience in businesses: the role of artificial intelligence. J. Cleaner Prod. 138228 (2023)

34. S. Singh, M.K. Goyal, S. Jha, Role of large-scale climate oscillations in precipitation extremes associated with atmospheric rivers: nonstationary framework. Hydrol. Sci. J. **68**(3), 395–411 (2023)

35. S. Singh, N. Kumar, M.K. Goyal, S. Jha, Relative influence of ENSO, IOD, and AMO over spatiotemporal variability of hydroclimatic extremes in Narmada basin, India. AQUA—Water Infrastruct. Ecosyst. Soc. (2023). https://doi.org/10.2166/aqua.2023.219

36. K. Toniolo, E. Masiero, M. Massaro, C. Bagnoli, Sustainable business models and artificial intelligence: opportunities and challenges 103–117 (2020)

37. N. Vivekanandan, S. Singh, M.K. Goyal, Comparison of probability distributions for extreme value analysis and predicting monthly rainfall pattern using Bayesian regularized ANN. 271–294 (2023)

38. J.A. Weyn, D.R. Durran, R. Caruana, improving data-driven global weather prediction using deep convolutional neural networks on a cubed sphere. J. Adv. Model. Earth Syst. **12**(9) (2020). https://doi.org/10.1029/2020MS002109

39. G.A. Wick, P.J. Neiman, F.M. Ralph, Description and validation of an automated objective technique for identification and characterization of the integrated water vapor signature of atmospheric rivers. IEEE Trans. Geosci. Remote Sens. **51**, 2166–2176 (2013). https://doi.org/10.1109/TGRS.2012.2211024

40. J. Zheng, X. Fu, G. Zhang, Research on exchange rate forecasting based on deep belief network. Neural Comput. Appl. **31**(1), 573–582 (2019). https://doi.org/10.1007/s00521-017-3039-z

Printed in the United States
by Baker & Taylor Publisher Services